LIVING WITH THE STARS

LIVING WITH THE STARS

How the Human Body is Connected to the Life
Cycles of the Earth, the Planets, and the Stars

*Karel Schrijver and
Iris Schrijver*

OXFORD
UNIVERSITY PRESS

OXFORD
UNIVERSITY PRESS

Great Clarendon Street, Oxford, OX2 6DP,
United Kingdom

Oxford University Press is a department of the University of Oxford.
It furthers the University's objective of excellence in research, scholarship,
and education by publishing worldwide. Oxford is a registered trade mark of
Oxford University Press in the UK and in certain other countries

© Karel Schrijver and Iris Schrijver 2015

The moral rights of the authors have been asserted

First published 2015
First published in paperback 2019
Impression: 5

Published in the United States of America by Oxford University Press
198 Madison Avenue, New York, NY 10016, United States of America

British Library Cataloguing in Publication Data
Data available

Library of Congress Cataloging in Publication Data
Data available

ISBN 978–0–19–872743–9 (Hbk.)
ISBN 978–0–19–883591–2 (Pbk.)

Printed and bound by
CPI Group (UK) Ltd, Croydon, CR0 4YY

This dewdrop world
is but a dewdrop world,
and yet—

    KOBAYASHI ISSA (1763–1827)

Preface

Our journey to "get the big picture" of the physical place of the transient human body within the ecology of the universe and of the links between the atoms in our body and the lives of stars, resulted in the book that is now in front of you. This book rests on a foundation of many conversations in which we talked about a multitude of topics from our everyday professional lives as a physician focusing on the wonders of DNA married to an astrophysicist exploring the secrets of the Sun and other stars. The more we followed up on our questions to each other, the more we discovered that our apparently disjoint fields of professional interest had far more connections than we had imagined possible. Eventually, our naïve questions to each other began to shape a web of links so fascinating that we decided to integrate it all into a full story that we could share.

The numbers involved in tracing the connections are simply staggering. Appropriately enough, we can call the number of stars in the Galaxy—at least 100 billion (and likely a few times that many)—astronomically large. There are, however, something like 500 times that many cells in the average human body: 50 trillion (and maybe twice that many). Each cell, on average, contains very approximately as many atoms as there are cells in the body. We can but marvel at the human enterprise called science that is rapidly advancing our understanding of this vast hierarchy of scales. Similarly, we can but be in awe of our own bodies, which, most of the time, manage to successfully run an assembly of individual cells that outnumber the human population of the entire planet by a factor of close to 10,000.

Even more than the matters of scale and the variety of the links to the world around us, it was the utterly transient character of our human bodies that struck a chord. We are not just taking in and burning fuel, like a car would, but instead we use our food to rebuild our bodies, over and over again throughout our lives. Very little of our physical bodies lasts for more than a few years, which is completely at odds with our feelings of continuity over a lifetime. It is that transience of the body, and the flow of energy and matter needed to counter it, that led us to explore the interconnectedness of the universe.

The research for this book allowed us to merge our fields and to become, at least for a while, "astrophysicians" in search of the natural flow that inexorably ties our human experience to the universe at large: the growing of a plant in sunlight, distant exploding stars, the cycling of cells in the human body, and the Sun's influences on Earth all are parts of our daily existence, and their consequences are continually flowing through our bodies even if we do not realize it.

We invite you to take a journey with us as we tour the vast network of processes that link us to all life on Earth as much as to our Galaxy. Before reading this book, you might think that these associations are farfetched, but we hope to have created a story that convincingly makes the case for these connections. We hope that you will share our fascination with life on this large but isolated rock in the emptiness of interstellar space that makes such remarkably efficient use of the omnipresent building material: stardust.

You will see just how ephemeral and transient our existence is, and that throughout our lives our physical existence is continually recycled and rebuilt. We truly live by the grace of the stars, both by the Sun, which is our nearest stellar neighbor, and by the many generations of stars that came before.

In the first two chapters, we set the stage and describe just how transient our bodily form is, even as our person survives for decades. In Chapters 3, 4, and 5 we discuss the chemicals going into the making and remaking of our bodies. The connections of these chemicals to the world around us are the centerpieces of Chapters 6 through 11. How the ancient elements were formed and how they came to shape our solar system are at the core of Chapters 12 and 13. In a short, final chapter we review the connections between our bodies and the world around us, and the links between the many sciences needed to uncover them. Together, these chapters form a complete story, best read in order. They can, however, also be read by themselves, as essays focusing on particular sets of topics that are parts of the whole.

The birth of this book would not have been possible without the help of numerous individuals to whom we are forever indebted. We thank our parents, who stimulated us to look around and follow our dreams. We owe a debt of gratitude to the inventors and developers of the internet and the creators of Wikipedia, who, combined, enabled us to find and read texts from around the world, some of which are centuries old while others were recently conceived, from explorers and scientists, including Nobel laureates and other giants of their times, and to the many scientists working in so many different fields on whose shoulders we all stand today to answer fascinating questions. So much information is literally at our fingertips in the search of knowledge. All we did was chase one link to another, find our way to where connections lay, and weave our story through it all. You will find no references in this text: the resources to which the internet gives you, the reader, access are so extensive that you should have no trouble at all looking up the original texts and background information going into this book and then some. We hope you will enjoy the incredible journey.

Contents

If we are
    stardust, if flecks of it glitter
        in our bones,

is some part of the stars
    our dust? Do bits
        of the dead—the unwrapped,

unembalmed, unfettered
    dead, those free of the trappings
        of immortality—do they rise

seventeen times as high
    as the moon? Has a strand
        of someone's red hair

threaded its way through
    Saturn's rings? Is dark matter
        mottled, here and there,

with cells from a bright
    blue eye? Do infinitesimal sperm
        swim out of the Milky Way

and toward Andromeda? Suppose
    some fragment of consciousness
        managed to land on Mars—of course

it wouldn't matter, without a mouth,
    an ear, without a soul
        to tell its story to ...
        SHARON BRYAN (1943–), excerpt from "Stardust"

The Illusion of Permanence

Our bodies are made of the burned-out embers of stars that were released into the Galaxy in massive explosions long before gravity pulled them together to form the Earth. These remnants now comprise essentially all the material in our bodies. Most of it has been cycling through the continents, with a small amount added only recently, when comets were captured by the Earth's gravity, or when ultrafast particles ran into the atmosphere and created a shower of new particles. It is being used by all living beings on the planet. Plants capture sunlight and release oxygen even as they create food for us, humans. We live by the grace of stardust assembled by plants into nutrients that provide us with energy to grow, to move, and to think. Those light-powered plant chemicals are used to rebuild our bodies over and over during our lives. This rebuilding is so common inside our bodies that every few years the bulk of our bodies is newly created, giving us continuity merely as a shape in which most of the tissues are repeatedly replaced.

When we are asked to define ourselves ("who are you?"), we can approach this challenge in various distinct ways. We can state that we are human and can add whether we are male or female. Moreover, we can comment on our physical features, list character traits, and include certain socio-economic factors, such as whether we are single or married, have children, are employed, and, if so, in what line of work, and at what level within an organization. Some may define themselves by their status in society, by their possessions, by their accomplishments, or by their interests and relationships. Whatever descriptions we choose, they will inevitably be rough and incomplete. They will also be influenced fundamentally by the culture and by the era into which we were born.

If, instead, we consider directly what actually constitutes us, we may think of the body and of the mind, with the two often being thought of as somewhat distinct entities. Some people may introduce the metaphysical concept of a soul or of a life force. Considering, for now, only our physical bodies, it is clear that we are made up of a hierarchy of organs, tissues, cells, and molecules. Our bodies are estimated to contain the amazing number of approximately 50 trillion human cells (using the so-called short scale for large numbers as we do throughout this book, so that means 50 million times a million cells). Within the cells that shape the human body, numerous tightly regulated processes occur to keep us alive and in a healthy state. These involve a large variety of molecules, which are composed of elemental atoms. Each cell contains an average number of atoms that is approximately as large as the number of cells in the body.

We may think of our bodies as composed of "our cells", but interestingly the majority of the cells that are contained within our overall human form are not human at all: on the inside and outside surfaces of our bodies, the bacterial cells that form what is called the human microbiome outnumber human cells by approximately 10 to 1. We are, in fact, colonized by hundreds of bacterial species. Bacterial cells are much smaller than human cells, so although they outnumber our human cells they do not considerably add to our weight. These bacteria are, however, critical factors in both health and disease because they influence complex biological processes, including the regulation of our immune system and the digestion of food. The immune system helps us ward off damaging effects of the surrounding world and digestion allows us to extract the nutrients that our bodies require from the food we eat. Thus, in order to survive we actually need this symbiosis that blurs the line between "us" and our surroundings.

Apart from the physical reality that is our body, another attribute by which we often define ourselves is our age. We rarely realize just how diverse a range of aspects we could consider in giving our age. Here, as in our thinking about what makes our physical form, considerable ambiguity arises in any attempt to formulate a straightforward number. Generally speaking, and as legally defined, our age is the time elapsed since birth. Yet we often allow for other evaluations. For example, there is the age of how we feel or of the degree of maturity with which we behave. Both are heavily influenced by the culture we live in. For example, in some places we may find many a grandmother well into her 70s who thinks nothing of going for a challenging hike each day, who dresses contemporarily, and who actively pursues a variety of interests. In contrast, we may encounter women of the same age in other cultural settings who have very sedentary lifestyles and who consider themselves as having all but lived out their lives. Not envisioning a compelling future, their behavior and presentation fully reflect that perception. The flip-side of this cultural difference is that old age is frequently resisted and rejected, especially in North American culture: age is something that is fought intensely, and weapons of questionable effectiveness are eagerly supplied by the beauty and biomedical industries. In some cultures, however, a ripe old age is bestowed with certain benefits, such as elder status and an inherent association with wisdom that demands respect. Thus, to some extent, age and its interpretation are mental and cultural concepts.

Another way to address the question of age is to consider our *functional* age, which is an index of an individual's performance, taking into account a combination of the chronologic, physiologic, mental, and emotional ages and is, as such, a useful indicator of aging or "senescence", more so than time since birth alone. However, it is only our chronologic age that we typically are expected to provide as the metric of how old we are when we are asked our age.

In reality, however, even the concept of chronologic age has an un-avoidable ambiguity to it. When not applied to a living being, chrono-logic age is generally taken to mean the length of time that has elapsed since the object was first completed. That works just fine for a painting, a piece of furniture, or a book. It quickly becomes less obvious, how-ever, for longer-lived objects, such as an old car with many replacement parts, a road that has been resurfaced or diverted several times over, or a centuries-old house with numerous additions and repairs.

So, how old are you? That deceptively simple question can be an-swered in very different ways. Nine months before humans are born, at the time of conception, the genetic complement of what that human will grow into is created when an egg cell and a spermatozoon merge to create the first cell of the new individual. Both the paternal and mater-nal components already existed separately before that, but their mer-ging is essentially the first moment when the blueprint of that specific human being, in strictly genetic terms, is present in the universe.

From the moment of conception onward, it takes some two decades to grow into what we consider to be an adult human. The moment of birth is clearly a differentiating moment during that time frame, and, if we were houses, we could think of ourselves perhaps as "finished", as ready to live in, after our development in the womb. During the entire growth period to adulthood, however, more material is added to our bodies than they contained at birth, and this accumulation of material does not stop when we reach the legal age that confers the status of adult. Even if we do not gain weight in later decades, which most of us do not manage to entirely avoid, the body does not cease to accumulate and process material from the world around us.

The chemicals that we take into our bodies when we eat, drink, and breathe do not simply enter and leave, but may become part of us temporarily, to help power our muscles and our brain, to repair cells, or to build new ones as earlier generations die. The original cell out of which we grew has, in all likelihood, long died. Only copies of it, differentiated to perform one of the hundreds of distinct functions within the body, continue to exist, variously lasting from days to decades, all to be replaced themselves at some time during our lives.

When we think of ourselves as a composite of cells with only temporary existence, it is evident that the conventional approach of listing our age as the number of years elapsed since birth utterly belies the complexity of the processes in our bodies. These processes reflect a sophisticated regeneration and recycling machinery that is, in turn, tightly connected to our terrestrial environment and even to our cosmic one.

We are continually rebuilt out of materials that stay in our bodies perhaps for only some days or maybe for a few years. But it goes further than that: some of the vegetables and fruits that we consume today contain energy trapped from the light that left the Sun only eight minutes before it was absorbed by a plant, and only weeks before that plant was harvested. The majority of the atoms that are present in our bodies have existed for as long as the universe has; others were forged inside stars billions (that is, thousands of millions) of years after the Big Bang, yet still billions of years in the past. Some atoms were made less than a human lifetime ago by nuclear reactions in the upper layers of the Earth's atmosphere, called the stratosphere, where commercial air traffic cruises along. What we document in this book is that this much is certain about us: very, very little in our bodies has actually been part of it for as long as what we say when we state our age. All the body parts of any living organism on the planet, be it plant or animal, are continually subject to replacement and recycling.

At some level, we are all aware of the reality of impermanence. We so often say that we no longer are who we were when we were younger. That is true enough in the psychological sense in which we generally mean it. It is particularly true, however, for the material that makes up our bodies, ranging from the water that constitutes the biggest volume of our bodies—being recycled on a time scale of days to weeks—to the very foundation of our bones and our nerve tissue, components of which may take a few decades to be largely replaced. What survives all this replacement is only a shape, a pattern, a temporary collection of matter and energy assembled into a host of specific, recognizable attributes that persist, at least for much of a lifetime.

The temporary collection of matter and energy that is a human body is a relatively flawless copy of what it was hours, days, and years ago. Even our genetic material, the DNA that lies coiled in the chromosomes in the vast majority of our cells, disappears when cells die, and is replaced in a new generation of cells. As years go by, the pattern of what we visibly are does change, yet it is still familiar enough for people to be able to recognize themselves in pictures of when they were children, and to recognize one another many years after a previous encounter. In such meetings, at high school reunions, at weddings or funerals, or in chance encounters in the street, we acknowledge that we have grown older (often perceiving more change in the other than in ourselves), but features and habits are sufficiently long-lasting for us to recognize each other, even though there may be some hesitation when much time has passed between successive encounters. Fortunately, many of our memories survive over the years, although it is unlikely that what contains these memories is composed of the same molecules that first captured them.

What it comes down to is that we are intrinsically impermanent, transient, continually rebuilt, and forever changing. We are a pattern, like a cloud, a traffic jam, or a city: even as we exchange our building blocks continually with our surroundings, the overall pattern provides enough stability for us to have a sense of continuity, both with regard to our own body and personality, and with regard to those of others.

Our bodily existence, in a way, is also akin to a wave in the ocean. The wave is a pattern of motion that travels in space and time, but the water that makes a wave at any given moment resides within it only briefly. The wave moves through, but not with, the water. Or, if we were to travel with the wave, the water would appear to flow through the wave, with as much coming into it on one side as is leaving it on the other. In the case of living organisms such as we are, the analogy with the water moving through the wave is imperfect because chemical elements and compounds stay in our bodies for different amounts of time. Nevertheless, the fundamental nature of the comparison holds true: we are a composite of patterns that are shaped by the chemical components coming into us and which travel with us temporarily before they are discarded and left behind as new ones are collected and incorporated.

We may not normally apply to ourselves directly this concept of existing only as a pattern, but with advancing technology we have become quite used to thinking in abstract terms of patterns and their evolution as copies of earlier versions. We put pictures of ourselves on social networks that friends can view on screens far away, with no idea of what was the original and what the copy. We enjoy duplicates of music files or videos on devices in our homes or on those we travel with. We email messages which, when printed, become copies that are not identical to the originals, yet carry the same content. Such persistent patterns long preceded technology and in fact surround us in the natural world every day. Let us look at a few examples.

A cloud is a pattern in which water condenses from vapor into droplets. As air moves slowly through the cloud, new droplets form while earlier ones may be lost by evaporation at the cloud's edges or by raining out of the cloud base. Another example is a river. Water moves in a river from the headwaters to the final basin, commonly a sea or an ocean. New water enters all the time, so that when we sit on the river's banks we are able to look at the one pattern that we designate "the river", but that pattern is filled with different water as time goes by and changes its shape. If we were living at a much slower pace, we would be able see the evolution in the pattern that we call a forest, too: individual trees sprout, mature, and die, yet unless that forest is cut down or burned completely, successive generations of trees form the continuing essence of this living forest. Speeding up time even more, we would see the mountains change into grains of sand as erosion takes its toll, with other mountains emerging on the Earth as a consequence of geological forces. On the longest time scales that fit within the existence of our planet, we would see continents move about, fracture, erode, and reshape, subject to the conveyor-belt motions of the seafloors. Even on the exceedingly long time scale on which stars evolve, a cyclic pattern is seen: all stars that are somewhat heavier than our own star, the Sun, explode at the end of their existence, throwing most of their matter back into the galaxies within which they formed, and new stars and planets are recycled out of that stardust to hold the pattern that is their galaxy.

The undeniable truths of impermanence and of the inadequacy of any unique definition of age apply to our environment as much as to us.

For instance, in some places the Earth itself is created anew and appears to be younger even than we are. On volcanically active islands like Hawaii one can stand on volcanic rocks that are no older than a year. One can also stand on a sandy beach that has grown markedly since the previous season. In both instances, the overall land mass is reshaped as material is moved around: from deep within the Earth to its surface on a volcanic island and from a shallow sea floor to just above sea level in the example of a beach. Volcanic rock forms by the cooling of liquid magma that has cycled inside the Earth possibly ever since our planet formed. The rock that materialized when this magma reached the Earth's surface may only be hours, days, or years old. In the case of a beach, the sand grains imply a story with more pieces to it. Once upon a time a piece of solid Earth was eroded into smaller fragments, which may eventually have been packed into sandstone and then fragmented once more, over a long period of time. This may have occurred while drifting on the ocean floor or perhaps in distant desert dunes, before finally—but just as temporarily—coming to rest on a beach that warmly supports us when we enjoy a tropical sunset. In scenarios such as these, it is easy to recognize that "age" can represent any of a variety of time scales.

The face of the Earth has changed dramatically over time, even though we might think that much of what we see has been and will be there forever. Life on Earth formed several billion years ago, but mammals have been here for only something like 60 million years, rising as the dinosaurs became extinct. Humans as a genus have existed for only a few million years. Cleary, many very substantive changes have occurred over this period, as some things were lost forever while others formed for the first time. If we focus on mankind during the past few centuries, for example, we realize that, whereas technology has changed our lives dramatically, and whereas there are definitely more humans now on the planet than 100 years ago, mankind itself is like the pattern of the wave: people are born, grow up and grow old, and eventually die, but as they come and go they flow through the pattern of mankind, which maintains a much longer and larger identity.

In the case of the universe, we typically think "age" less than "ageless". Our eyes view the stars over a negligibly small time span from the perspective of our ancient universe. To them, the Galaxy is a vast, still, and steady cradle for our small planet. We readily observe the cycles of the Moon and we unavoidably live with the daily rising and setting of the Sun. Apart from these familiar motions, however, nothing seems to

change. If we measure precisely enough, we can establish that the stars
and the planets are moving through the firmament, but on the cursory
inspection that we commonly afford the skies, the universe appears to
us to be static. We usually do not stop to think that some of the stars
that twinkle in the midnight sky were extinguished eons ago, their light
having left the distant star perhaps thousands of years before that, still
traveling toward us as the star itself exploded and vanished. Despite the
universe's appearance as a timeless, permanent fixture, individual objects in space have different ages too. We can trace back the universe as
a whole to moments after the Big Bang, but most of the stars, planets,
and other heavenly bodies that make up the uncountable galaxies have
not existed since the beginning of time. Instead, early generations have
"died" while new ones have been formed.

Realizing how fluid we are and how impermanent everything around
us truly is, we should think differently about the materials that we take
into our bodies to quench our thirst and feed our hunger: this material
does not simply flow through our gut, but is assimilated into the very
structure of our bodies until, after some time, it leaves it again in some
other form. Our bodies are very complex chemical machineries, as are
those of all living things on this planet, but the mixture of the elements
used in our chemistry simply reflects what is readily available around
us in a chemically useful form and quantity. From this, it follows quite
logically that the most common elements in the planetary system are
the same as the most common elements in our bodies, albeit with some
notable exceptions that we can easily understand.

Not only the elements in our bodies, but all processes on the Earth are
directly connected to those in our solar system, to our Galaxy, and to
the universe beyond. From that perspective, it becomes understandable
not merely that life on Earth reflects the elemental makeup of the solar
system, but that the solar system in turn reflects the elemental makeup
of the Galaxy around it. Once mankind learned the true nature of stars
and realized that they are vast nuclear furnaces that shape and destroy
the planets around them, and when a better understanding emerged of
the central role of stars in establishing the contents of the universe, it
became clear that we are, indeed, stardust, in a very literal sense. Every
object in the wider universe, everything around us, and everything we
are, originated from stardust. Thus, we are not merely connected to the
universe in some distant sense: stardust from the universe is actually
flowing through us on a daily basis, and it rebuilds the stars and planets

throughout the universe as much as it does our bodies, over and over again. In our everyday lives, we tend to ignore the universe beyond the Earth's atmosphere and take it for granted that the Sun steadily shines its warming light onto the planet. We do not generally ponder the many links between us and the stars, except perhaps for the links to the nearest star that we call the Sun. These links range from the Big Bang, in which the universe was formed, to the particle radiation from the Galaxy that slams into the high terrestrial stratosphere right now.

With the exception of hydrogen, which formed when the universe itself was created, almost all of the atoms in our bodies have been forged either in the deep interior of stars, or within the explosion of these stars at the ends of their existence.

One key connection between ourselves and the universe and, indeed, an essential ingredient of life, has yet to be discussed: energy. It is energy that maintains the surface of the Earth well above the cold of the ambient universe, which is a mere three degrees above absolute zero and thus hundreds of degrees below freezing. In units that we encounter more frequently, this translates into about minus 455 degrees Fahrenheit or minus 270 degrees Celsius. Without an external energy supply, life on Earth as we know it would be simply impossible. The external energy raises the Earth's temperature so that it can support plants, which, in turn, feed animals, including us.

The bulk of the energy that is available to life on our planet comes from the Sun. Only a very small fraction of the Earth's overall energy budget is liberated from the radioactive decay of stardust in the Earth's interior. This afterglow of stellar deaths deep inside the Earth manifests itself to us on the surface as geothermal or volcanic energy. This energy is primarily liberated from the decay of uranium, thorium, and potassium, which were formed as stars exploded at the ends of their lives.

Even the Sun's radiant energy is a form of stardust. The energy of the Sun's light is ultimately generated in the deep solar interior by violent nuclear collisions. Mass and energy are equivalent and exchangeable: the solar light that hits the Earth loads the equivalent of about 140,000 tons of mass onto our planet every year. Much of that load is there only briefly, because a lot of the sunlight shining down on the Earth is reflected back into space by the clouds, oceans, and land masses, but part of it is stored in plants for days or weeks and then can be processed in a great diversity of ways. It may be used to build the compounds that constitute our bodies and to power the network of metabolic reactions that we need every single moment of our lives. Or it may end up stored underground, in plant and animal material that is slowly being converted into the chemical energy stored in coal, gas, and oil.

These connections between us and the universe offer us a view of the richness and diversity of the sciences that we know as chemistry and physics, as astronomy and biology, and as geology and medical science. Understanding them in detail is daunting, but exploring them broadly—as we do in this book—enables us to see ourselves as part of the greater whole. It shows us to be entities that do not exist apart from the world around us in a hostile universe, but rather as connected to all life around us, to the planet as a whole, to the Sun, and to the stars throughout the Galaxy. To explore this multitude of connections, we travel through this general hierarchy of scales in space and in time: from cells to atoms through the biosphere, from there into the cycles of the Earth, and then to the lives of the stars, to eventually return from the Galaxy to our solar system and the diversity of its components.

Key Points: Chapter 1

- The human form comprises some 50 trillion cells. It hosts another 10 times as many bacteria, many of which are integral parts of our digestive and defense systems.

- Human bodies are intrinsically impermanent. Rather than fixed, they are more akin to a pattern or a process, although stable enough to allow a perception of continuity.

- The chemicals that we take in by eating, drinking, and breathing typically become part of us temporarily to help power our bodies, repair our tissues, and replace cells. All parts of any organism on the planet are continuously subject to replacement and recycling.

- Not just our bodies and other living organisms, but all processes on the Earth are intertwined with those in our solar system, our Galaxy, and the greater universe.

- Our chronological age does not reflect the time during which the components of our bodies have been part of them. This is also true for our environment and the universe. Although the universe appears timeless, stars and planets come and go, and therefore have a range in age, too.

- Every object in the universe, everything around us, and everything we are originated from stardust. Stardust continually flows through us, directly connecting us to the universe, rebuilding our bodies over and over again.

- **The universe is constantly in flux. Everything is impermanent and interconnected.**

Each living creature must be looked at as a microcosm—a little universe, formed of a host of self-propagating organisms, inconceivably minute and as numerous as the stars in heaven.

CHARLES DARWIN (1809–82),
in *The variation of animals and plants under domestication, Vol. II* (1868)

The existence of forgetting has not yet been proven; all we know is that we are not capable of recollection.

FRIEDRICH NIETZSCHE (1844–1900),
in *Morgenröte; Gedanken über die moralischen Vorurteile* (1881)

Dying to Live

The human body performs a great variety of critical functions to keep itself alive. It does so to varying degrees of ability and comfort throughout our lives, the extent of which primarily depend on our age (here defined as time since birth) and the quality of our health. Most of the functions of the human body escape our conscious awareness, so that we generally do not notice that behind the scenes our body is a beehive of activity. We also fail to recognize that, in the process of keeping us alive, our cells divide, age, grow, and die in such a way that our body is a continually changing composite of which little survives over the years, with only copies of copies of copies of cells carrying us from conception to old age.

Take the skin, for example: a living, breathing, regenerating tissue that is the largest organ of the body and that acts as a barrier between the internal organs and the environment. In adults, it encompasses about 22 square feet (2 m²) and weighs around eight pounds (4 kg). It protects the interior of the body from injury, from harmful effects of microorganisms, and from the damaging ultraviolet rays of the Sun. It plays a role in the body's thermal regulation through the constriction or dilatation of small blood vessels, it contains nerve endings that allow us to feel touch, temperature, pain, pressure, and vibration, and it slows the loss of fluids from the body. The skin also shelters the hair follicles, which produce the hairs that cover most of the body's surface, and it provides storage for a variety of substances.

The skin, composed of several layers, ages quickly but is remarkably effective at renewing itself. In the top layer, the epidermis, most cells eventually reach the surface as the outermost layers of cells wear off. They are replaced in a time frame of roughly a month or two, in a continuous process that culminates in the loss of approximately 30,000 cells every minute throughout our lives. This translates into roughly eight pounds (4 kg) of dead material per year. Some features of our skin are, of course, more lasting. For example, we may have seemingly permanent moles and we may have scars that persist for years. These tissues, however, are not really skin. Moles are embedded within our skin but they are in fact benign growths that are typically composed of pigmented cells that do not follow the same lifecycle as true skin cells. Likewise, scars are repairs of deep cuts in our skin, but their fibrous repair tissue is organized differently than true skin, in that it lacks the typical characteristics of skin anatomy.

Another example of a cell type with a high rate of replacement is the blood cell. This type comes in two main classes. The red blood cells carry oxygen, while the white blood cells are important components of our immune system. Both cell types are predominantly produced in the bone marrow, with additional manufacturing capabilities in a few other organ systems, mainly the liver and the spleen. Subsets of blood cells are generated at the amazing pace of some 300 million per minute (this number, as many that follow, is a rough estimate and such numbers are subject to sometimes large uncertainties, while they also differ from individual to individual and from time to time). This massive output is required because of the equally high death rate of these cells. Red blood cells, much more numerous than the white blood cells, have a rather short life span: they mature in the bone marrow for a week, enter into the blood stream, and subsequently circulate in the peripheral blood stream for only four months before being trapped and destroyed in the spleen, with eventual recycling of their oxygen-carrying iron content.

Epidermal skin cells and red blood cells are just two examples of our many different cell types. In fact, every tissue turns over its cells, each at a different rate. This has been investigated using carbon-14 dating (to which we return in Chapter 8). When this dating method is applied to the DNA molecules that store our genetic identity, the life span of individual tissues in our bodies can be determined. With this technique, it has been estimated that cells in the adult human body have an average age of 7–10 years, which is far less than the average age of a human.

There are remarkable differences in tissue ages, however. Some cells exist for only a few days. These tend to be the ones that form the part of our body that is directly touched by the external world. This part is a constellation of external surfaces made up by our skin, as well as by the internal surfaces of our lungs and the digestive tract, which goes all the way from the mouth via the stomach through the gut. We are, in a simplified view, after all comparable to a doughnut or to a cored apple in which the outside world touches what we see as the external surfaces as well as what we do not generally see and then refer to as internal surfaces.

One inner part of our interface with the environment, the lining of the gut, is highly exposed to material from outside and consequently cell turnover there is correspondingly high. This rapid replacement of cells helps maintain a surface layer of cells that slough off rapidly but that collectively are essential in sustaining the physical barrier that protects us against harmful influences. Cells deeper in the intestinal tissue, which are not directly exposed, in contrast, may last upwards of 16 years.

The epithelial cells that pave the surface of the lung may exist for up to half a year before they are replaced, and some of the liver cells persist for a year or two. The muscle cells that constitute the heart typically continue to function for several decades; by the age of 50, about half of our heart cells have been replaced. Because of such substantial differences in the life span of the cells in the different tissues throughout our bodies we cannot assign an accurate age to our body, but it will be clear from the above that it takes at most a few years for most of the cells to be completely replaced, leaving in essence a copy of our "original selves" to go through successive cycles of replacement.

Because of the replacement of the contents of our cells and of the cells as a whole, in some sense we can make the case that we are always merely days, months, or years old, depending on which parts of the body we choose to focus on. The key concept is that our bodies are never stagnant and that our cells have to remain dynamic in order to stay alive. We have hundreds of different cell types, each with its own specific rate of cell turnover. Paradoxically, all this exchange and renewal serves to support what we think of as a rather persistent physical body.

All these examples of cell replacement make it clear that the vast majority of our cells are considerably younger than our chronological age. We tend to assume that nerve cells, particularly those in the brain, are with us for life after they have matured during the growth phase in childhood. This is indeed the case for some cells within the nervous system. One example is tissue in the visual cortex, the part of the brain that processes visual information and that enables us to realize what our eyes see. Carbon-14 dating has revealed that the age of these cells does indeed match our chronological age. However, findings in individual cell types should not be generalized to all cells in the nervous system. The question of how long a long-living brain cell in fact survives intact turns out to be a question of definition and remains very difficult to answer, because cells are made up of many components that are themselves continually replaced and rebuilt. For example, just the replacement of all the water in our bodies on a relatively short time scale of at most a few weeks demonstrates that the liquid contents of a cell change in a far shorter period than the lifetime of a cell, which is generally already considerably shorter than a human lifetime. Apart from water, we also need fuel to make our cells run. Brain cells, for example, thrive on the energy derived from oxidizing sugar, which consequently requires constant replenishment. Even the cell membranes, which package the cells, are maintained and replaced as needed, like fences that need mending from time to time. All these replacements occur gradually and continually. It is, therefore, difficult to accurately establish how long the entire replacement process typically takes.

Memories are also subject to impermanence: some short-term ones may never transform into long-term ones, and even those that we retain for more than a brief moment are not necessarily stable, but can be lost over time or disrupted by injury. Take, for example, the case of Henry Molaison, who has become the most intensively studied patient with memory loss in history. By the time of his death at age 82 in 2008, more than 100 researchers had studied him. He had brain damage that caused the inability to form new memories. This memory loss, called anterograde amnesia, was not for past events but for anything that happened after the brain damage. Thus, it was as if he were always experiencing almost everything for the first time. This severe condition was the unexpected side-effect of an operation done in 1953 during which neurosurgeon William Scoville removed certain parts of the brain that were thought to harbor the triggers of the grave epilepsy from which Henry Molaison was suffering. That surgery was performed because, at the time, there was no way to relieve these symptoms with medications. Although the surgery much improved the otherwise intractable epilepsy, it also resulted in profound memory loss for "H.M.", as he was called in the scientific literature in an attempt to protect his privacy. Doctors William Scoville from Connecticut and Wilder Penfield, a cognitive neuroscience professor from Quebec, started to record and study the memory loss, and thereby formed the basis of an entirely new understanding of the intricacies of our brain and memory system. Henry Molaison, by so generously participating in numerous tests and studies, through his tragedy tremendously helped to advance brain science.

To date, no scientist has been able to capture a memory, extract it, and store it as one could an object. Yet, over the past few decades, tremendous strides have been made in our understanding of the different types of memory, the anatomic locations that are important in memory creation and storage, and how memories transition from short-term phenomena to more permanently saved forms. Memories may not be solid objects, but many are lasting and all have a biological basis in our brain. In other words, there are processes in place that maintain our memories over time, even though their biological components are subject to change as well as replacement. This is an area of very active research in which multiple hypotheses about the workings of our memory systems are being tested. Some of this work builds on previous concepts, while other work opens entirely new avenues of thinking about memory. We just do not really know yet how memory works. Nevertheless, it is fascinating to think that something as personal and ephemeral as a memory has an age and a life span, fairly analogous to what we observe with the cells in our body.

Memories can be divided into two major categories. Declarative memory reflects our common meaning of the word "memory", which includes awareness of events and facts from the past. The second category of memory covers actions or procedures learned in response to what is presented to us by our environment. Such learned actions can become quite habitual and are typically not consciously recognized but instead are stored as a more internalized memory that is, nevertheless, based on previous occurrences. It is especially this procedural, implicit type of memory that is vital for our survival.

Through observations and studies of people and animals with damage to different parts of their brains we know much about the roles and functions of such damaged areas. For example, the area of the brain that is called the hippocampus (located in the center of the skull, just above the plane of the eyes and ears) is now known to be critical for the formation and retrieval of memories. Areas of the temporal lobes on both sides of the hippocampus further shape and maintain memories. Diverse other areas of the cortex, the outer layer of our brain, are also important for memory storage and recollection. There is, then, no single "memory control-center". In fact, multiple different areas of the brain are activated when we recall complex events, such as a first night at a performance with colorful costumes and beautiful music.

Memories do undeniably have connections to particular structural components of the brain but in fact they may best be thought of as a collection of patterns rather than as separable solid objects. When viewed this way, our overall memory can be described as a system of different circuits within the brain that are updated, influenced, and regulated by our environment, which may be stimulating the system or slowing it down. External factors that affect this system include drugs, caffeine, and the amount of sleep we have had. Importantly, the memory circuits are also controlled by internal substances such as neurotransmitters that help convey messages from neuron to neuron in our brain. Neurotransmitters, in turn, are regulated by changes in gene expression and other genetic switches, as well as by what are called epigenetic influences. These are not directly obvious in our DNA but can have powerful functional effects on our genes and their products. All of these components together help us to create the initial memory and enable us to subsequently consolidate, store, and retrieve it under the influence of newly synthesized proteins and neural pathway connections.

Our brain is thus not a static organ that does not change after the initial phase of growth and development, but rather contains networks that are ever changing, being rearranged and remodeled based upon external and internal environmental influences. The synapses, which are a critical part of our neurons that facilitate communications between the cells of the nervous system, are subject to change in their shape and in their direct molecular environment. Synapses may also change in number and can be added to some areas of the brain when needed, even in adulthood, to create networks that have resilience and some redundancy. This helps to create a more lasting trace when a memory is consolidated and stored. Neuronal pathways and their communications shape our image of the world in which we live, our habits, and our persona. Somehow, memories often manage to survive, albeit subject to gradual change, even as their substrates, the cells and their chemicals that make up our bodies, are replaced piece by piece over time.

The process of cell turnover continues throughout our bodies for as long as we live, beginning in the embryonic stage as a normal part of growth and development. In fact, in these early stages, much cell death is programmed into our DNA. For a developing embryo the planned death of cells enables, for example, the separation of our fingers during limb development, as well as the proper formation of the reproductive organs when they differentiate into male or female. On a daily basis, however, even as adults, we lose billions of cells through cell death that is designed into the human system. Amazingly, every year of our lives we each seem to use, eliminate, and regenerate a cell mass that is almost as large as our body weight.

The replacement of old generations of cells by new ones is fundamental to our health: cells die in order for the body to thrive. It is an elaborate chain of events that occurs in human tissues under normal circumstances according to a programmed interplay of cell division and cell death. Dictated by this tightly regulated process of cellular proliferation on the one hand and cellular attrition on the other, cell turnover amounts to a form of self-renewal, as long as cell creation and death are balanced. This self-renewal ensures a stable, well-adjusted environment in our tissues through the purging of some cells, cell division to replace the eradicated cells, or formation of new cells that specialize their function so that they can take on the roles needed for specific tissues. Exactly how a cell is slated to die and how cells are activated to divide to produce new cells, and also what sets the rate and quantity of these actions, are scientific questions that are not yet fully answered.

In some (and perhaps in all) cases, the cells that are being eliminated somehow communicate with the surrounding cells to signal the need for an environment within that tissue that will stimulate cell division to replace dying cells. This cell division can occur in a variety of ways. When mature, specialized cells divide, they essentially make duplicates of themselves. In many instances though, cell division does not primarily originate from cells with particular functions but rather from what are called "adult stem cells" or from their descendants, which are called precursor cells. Stem cells are not-yet-specialized cells, or undifferentiated cells, that are not programmed for a specific cell-type destiny. Such cells are present in virtually all of the human organ tissues. They are called into action when needed to support the cellular environment in which they are located. Upon division, the newly created cells specialize as the specific types of cells that had to be replaced.

The notion that there are undifferentiated, unspecialized stem cells dates back just over a century, but the first description of the multipotency of these cells in normal tissues was not until 1963, by biophysicist James Till and experimental hematologist Ernest McCulloch, from Toronto. They had earlier started to collaborate on a project that aimed to find out how the human body could be protected against the effects of the atomic bomb, the threat of which occupied the minds of many at the time of the Cold War. In the course of their work they exposed mice to lethally large amounts of radiation and discovered that certain cells were capable of differentiating into any of the various types of cells we find in the blood, rather than just being programmed to follow a single predestined path. In addition, they figured out that these stem cells were capable of self-renewal. Their discoveries paved the way for stem cell research well beyond the blood system and it remains a very active area of medical research today.

Although the details of the integration of cell death and division in adult tissues remain incompletely known, it is clear that effective management of this process requires a complex series of actions that involve much more than small clusters of cells here and there, and that therefore must involve cell signaling pathways that control the larger tissue environment and possibly entire organs. The number of genes within our DNA that have been shown to be involved in coordinating this balancing act between cell birth and death continues to increase. More and more components of this process are gradually being revealed through discoveries that include interactions within the various cellular signaling pathways, and the regulation of messages that are being exchanged within cells and between cells.

Cell replacement is not only important in maintaining our body over time under normal conditions; it is also critical when tissues are in need of repair following damage through disease or injury. As we advance in age, however, our cells have a diminished capacity for self-renewal and a reduced capability of tissue regeneration. The process that is at the center of aging and that ultimately tips the balance toward cell loss is called biological aging or cellular senescence. Cellular senescence reflects a state in which cells are no longer able to adequately divide either to meet the demands of normal cell turnover or in response to damage. In this condition, an important source of tissue renewal is gradually lost: when the pool of stem cells and their offspring precursor cells fail to maintain their regenerative fitness, the very source of rejuvenation is depleted from our tissues. Interestingly, although the lack of cell division associated with cellular aging overall is detrimental, it can also be protective when it precludes the proliferation of cells that have undergone unfavorable changes that could eventually make them cancerous. Unfortunately, this mechanism of protection is only one factor in a myriad of processes: in reality, overall there is an increase in the frequency of cancer development with advancing age.

Tissues lose their regenerative potential over time as the number of senescent cells increases and the physical decline that accompanies the aging process becomes obvious. Aging, however, is not solely determined by senescent cells, but also by a general loss of function in those cells that are fully specialized and therefore no longer dividing, but that still remain capable of living. Such a loss of overall function may be tied to an impaired ability to communicate with other cells or to problems with the synchronization of a variety of cellular tasks. Moreover, aging cells release substances that influence critical cellular processes. These processes include cell growth and differentiation, the creation of new blood vessels after injury, and tissue structuring. If inappropriate chemical signals are released from senescent cells, the structure and function of other cells can be disturbed. A general response to that unfavorable state is a chronic inflammation that itself further upsets the functioning of the tissue. This type of inflammation is not caused by bacteria or by viruses, but is a negative consequence of the chemicals released by aging cells.

In addition to internal tissue dynamics, the process of aging is also considerably influenced by the environment as well as by a person's lifestyle. These effects can either accelerate or slow the effects of aging. Even under the best of circumstances, however, the rate of overall aging rapidly increases with the number of years since birth. In the end, we do not simply die of old age, as it is often said, but rather of a decline in cellular vitality, which in turn permits diseases to develop when cell damage accumulates and prevents cells from being efficiently replaced. Nevertheless, given that we do not seem to all age at the same rate, there must be individual differences. The rate at which we age and that at which age-related diseases manifest themselves are moderated by internal reactions. These reactions regulate the body's healing and the body's response to conditions that cause stress to our tissues and cells. Such conditions include internal challenges, such as process errors that cause damage to our DNA and the malfunctioning of proteins. And they include external influences, such as oxygen availability, cell nutrition, temperature, physical activity, and exposure to unhealthy chemicals and radiation.

External influences are difficult to study in a standardized way over a lifetime and new findings about the decline of overall health with advancing age have therefore often resulted in scientific or political controversy. Geneticist Edward Lewis, for example, performed groundbreaking studies of genes that play a role in the controlled development of each body segment in the fruit fly. The scientific insights he obtained from this tiny insect turned out to be widely applicable to human development as well, so much so that he received the Nobel Prize in medicine for his work in 1995. Lewis also had another interest, however. In the 1950s he studied the effects of ionizing radiation, both from nuclear bombs and from X-rays, on the human body. It had previously been known that high doses of radiation could cause diseases and mutations in genes, but at the same time there had been the belief that low doses would be safe. Lewis publicly challenged that notion and argued that the development of leukemia and other cancers does not happen only when a certain exposure threshold is exceeded in a particular instance, but that such conditions can be caused by the sum of all doses of radiation over time, even when low. In other words, every dose of radiation carries the risk of damage to our DNA. He published this work and presented it to an atomic energy committee of the U.S. Congress. There was vigorous debate surrounding his findings, but in the end his research on the health hazards of radiation raised public awareness and was instrumental in the creation of radiation safety standards.

Research has revealed that it is possible to manipulate the cell signaling pathways that mitigate stressful conditions in the cell. If this could be done in humans in a safe, effective, and individualized way, then the stress resistance of our bodies could be improved. That, in turn, could translate into a longer personal health span, fewer age-related diseases, and new therapies against aging.

Key Points: Chapter 2

- The cells in our bodies remain active throughout our lifetime. Numerous tightly coupled cellular processes keep the body functional, healthy, and alive.
- Our bodies are composed of hundreds of different cell types, which have a wide range of natural life spans.
- The process of cell turnover is critical to our health: cells die and new ones are generated so the body can thrive. This process takes place in all human tissues under normal circumstances, based on a programmed interplay of cell division and cell death.

- Cells divide in order to maintain tissues as other cells die. Many specialized, differentiated cells do this, but often this replacement by division involves stem cells embedded within the tissue, of which only the offspring later differentiates to turn into a particular cell type with specific functions.

- **We purge and regenerate a cell mass that is about as large as our body weight every year of our lives, involving millions of cells each second.**

All would live long, but none would be old.

BENJAMIN FRANKLIN (1706–90),
in *Poor Richard's Almanac* (1749)

When I was younger, I could remember anything, whether it had happened or not; but my faculties are decaying now and soon I shall be so I cannot remember any but the things that never happened.

SAMUEL L. CLEMENS (1835–1910),
in *My Autobiography: "Chapters" from the North American Review*, by Mark Twain (1907)

Countering Wear and Tear

Advanced age increases the risk of getting a variety of human diseases. Some of these conditions are due to the regular wear and tear on the body, similar to the failing of the mechanical parts of a tool or a vehicle. The most common joint condition, osteoarthritis, for example, falls into that category. This is a degenerative joint disease that in most cases appears in middle-aged individuals and is the result of the wearing off of the cushioning cartilage between our bones, which leads to stiffness, pain, and swelling. Apart from the wear and tear of the cartilage itself, there also is a gradual stiffening of the surrounding ligaments and muscles, especially noticeable in the morning, and it can become harder to move overall. Although some people never develop complaints, everyone of advanced age has some evidence of osteoarthritis in their joints. Along the same lines, age is also associated with a tendency to lose muscle mass, with diminished functionality of organs (including the liver and kidneys), and with the loss of vision and hearing. In all of these losses of functionality, the tissue environment is altered with age and the cells in these tissues function abnormally.

Another category of age-related diseases is that of the neurodegenerative conditions, such as Alzheimer's disease. This form of dementia is associated with changes in memory, in behavior, and in the capability to think, speak, and understand. The hallmark of neurodegenerative conditions is that symptoms worsen over time. Although Alzheimer's disease is age-related and therefore more frequent in old people, it is different from osteoarthritis in that it does not affect everyone who simply reaches a high enough age. It is a specific, currently incurable disease that seems to be influenced by hereditary as well as environmental factors. Nevertheless, our brain is subject to general decline as much as the other parts of the body, and, although not universal, some forgetfulness and slowing of the thinking process is very common in the elderly.

Chronic diseases such as type-2 diabetes are also more commonly seen with advancing age. The tissue cells of someone with this condition cannot properly respond to the hormone (which is a chemical used by the body to achieve long-range cellular effects) known as insulin, which is produced in the pancreas. Insulin is needed to enable the uptake of glucose from the blood stream by cells and insufficient insulin causes blood sugar levels to become too high. That, in turn, can lead to early symptoms such as fatigue, increased thirst and urination, and blurry vision. More serious complications of elevated blood sugar levels that develop over time can include heart disease, loss of eyesight, trouble with feeling due to nerve damage, and kidney failure. Although the development of type-2 diabetes has much to do with heredity, there also are lifestyle-related issues that have a major influence on the development of this disease, including obesity, lack of physical activity, and an unhealthy diet. Basically, it comes down to the fact that the excess body fat that is a result of being overweight interferes with the function of insulin, so that blood sugar cannot be stored inside the cells as a source of energy. Given that lifestyle choices play a key role in the development of this serious disease, changes in diet, exercise, and weight can vastly improve the condition, and in some cases completely reverse it. Age alone, therefore, is not the essential determinant of this chronic disease.

Diseases that affect the heart and blood vessels, like atherosclerosis, are other examples of conditions that become more prevalent with age. Atherosclerosis is characterized by a hard build-up in the arteries that, among other things, includes cholesterol. This causes a narrowing of the blood vessels that renders them less flexible and impairs normal blood flow, which in turn deprives tissues of oxygen. Ultimately, blood flow locally can be blocked altogether, with catastrophic consequences, such as a massive stroke or a heart attack. As in the case of type-2 diabetes, atherosclerosis, too, emerges after the continued presence of frequently preventable detrimental effects related to a person's lifestyle. For atherosclerosis these effects include a high-fat diet, lack of physical activity, being overweight, abnormal lipid values in the blood, smoking, high blood pressure, diabetes, and heavy alcohol consumption. This is not to say that someone with excellent health habits will not develop some atherosclerosis with increasing age, but the rate of accumulation of harmful substances in our vessel walls can be slowed down tremendously by a healthy lifestyle.

All these examples illustrate that age is a significant factor in the development of diseases, some of which are due to wear and tear, some of which develop in only a subset of people, and some of which are a combination of aging and the consequences of unhealthy lifestyles. When deterioration persistently outpaces repair, cellular loss afflicts organ function and ultimately causes organ failure. In a sense, therefore, advancing age is a chronic degenerative condition, culminating in multisystem failure and, inevitably, death.

Throughout history, humans have sought to slow the aging process in a remarkable number of ways. Many attempts to preserve our youth are rooted solely in lore but some strategies are based on scientifically verifiable concepts. It is not surprising that a healthy lifestyle and a limited but sufficient calorie intake increase the probability of a healthy old age, whereas unhealthy states such as addiction and obesity typically accelerate the effects of aging. The biology of aging, however, is multifaceted and incredibly complex, with much yet to be discovered, verified, and connected. One research effort in this area is the sequencing of the entirety of the hereditary information (the genomes) of people of the age of 100 or above, from all around the world. By sequencing the genome, which in humans consists of a staggering 6 billion building blocks (called base pairs) that determine our traits, information from our approximately 25,000 genes can be derived and analyzed.

Such extensive genetic analysis was not possible until recently. It was only in 1953 that James Watson, Francis Crick, Maurice Wilkins, and Rosalind Franklin figured out the structure of our DNA. Watson, Crick, and Wilkins shared a 1962 Nobel Prize for this work. Franklin died from ovarian cancer in 1958 at the age of 37. Her scientific contribution is often overlooked, but her X-ray diffraction images of the DNA were absolutely central to the discovery of the double-helix structure in which one chemical strand spirals around its partnering one.

Once DNA's three-dimensional shape was known, genetic research accelerated, leading to the discovery of many genes in humans and in other species, and in the ability to identify those DNA changes that cause disease. The 1990s saw the beginning of an international research effort to sequence the entire human genome. A rough draft map of the human genome was finished in 2000, presented jointly by Francis Collins and Craig Venter, who were originally fierce competitors in the race to the finished genome. The International Human Genome Sequencing Consortium declared the human genome project essentially completed in 2003.

With this milestone, we now have the framework to study the genome as a whole instead of looking at only one gene at a time. Rapid technological improvements in sequencing speed and accuracy together with a continuing decrease in cost for this new ability give tremendous potential to genetic research. It holds the promise of what is called "precision medicine". In other words, when genome analysis can be incorporated into the medical care of each individual patient, it may help to determine that person's inherited risk of developing certain diseases and could translate into personalized recommendations for the prevention of disease; it could even assist in the delay of aging. In addition, such comprehensive genome-level testing by molecular pathologists and geneticists could improve the diagnosis of diseases. Once a diagnosis is supported by a detailed genomic profile, it may then be possible to make predictions of the course and severity of a disease. Finally, by having knowledge of how genome variants influence the body's response to pharmaceutical treatments, it might become possible to provide a patient with the best medicine at the right dose and at the right time, all tailored to that individual. By sequencing the genome we get an unprecedented peek into the molecular basis of human variation, of health and of disease. One of the aims of sequencing the genomes of centenarians is to hunt for those genes and other genetic elements that protect against aging and that promote a long life span.

For an adult body to remain healthy there should be at least a rough balance between cell death and cell creation. With some 50 trillion cells in the body, with typical ages of days to a decade, every second of our lives millions of cells die and the same number is newly created. Maintaining such a balance is a task of astonishing complexity, which the human body, most of the time, manages without us noticing significant problems. An excess of cell death over cell creation leads to physical decline. The balance may swing the other way, when unrestrained cell proliferation occurs. This is the keystone of cancer, which can present itself in the blood as a form of leukemia or in tissues as a solid tumor.

We now know that the trigger for uncontrolled, cancerous cell proliferation lies in changes in the DNA. Originally, however, it was thought that cancer did not result from DNA abnormalities but rather that it preceded them. In the early 1970s, Janet Rowley at the University of Chicago turned that notion upside down. She was a physician-scientist who, with four sons to take care of, initially worked as a part-time doctor. She fell in love with research only in her 40s, when she started her work on human chromosomes. With the limited techniques available back then, chromosomes were quite difficult to analyze. Nevertheless, she identified the first known rearrangement of segments of genetic material between chromosomes (known as a chromosomal translocation) in a form of leukemia, soon to be followed by more such discoveries. Each of these is associated with one or more specific forms of the cancer groups of leukemias and lymphomas. One such abnormality entails the exchange of genetic material between chromosomes 9 and 22. In both these chromosomes, a piece is broken off and joined onto the other chromosome, so there is no loss of chromosome material overall, but some of it is now located in the wrong place, thereby thwarting the normal function of the affected genes. Her discoveries transformed our understanding of cancer and were an important step in the development of targeted cancer therapies. Blood cancers caused by the exchange of DNA segments between chromosomes 9 and 22 are now among the most successfully treated cancers.

Cancer is associated with uncontrolled cell division, and, in time, cancer tissue can grow to invade neighboring tissues and blood vessels. Once blood vessels are breached, this may lead to the transport of loose tumor cells to other areas of the body, where they can settle, continue to proliferate, and develop into distant tumors of the original cell type; these distant tumors are called metastases. The uninhibited cell proliferation at the root of cancer is precisely the opposite of cellular aging, which leads to a reduced potential for cell generation, so it may seem surprising that malignant diseases are connected with advancing age. However, this paradox is explained by the fact that the chemicals that are secreted by senescent cells have an effect on their neighboring cells and on the surroundings in which those are situated. Some of these chemicals derail normal cell division, promote the development of tumors, or stimulate the invasion of neighboring tissues by cancerous growths.

In the United States, deaths caused by cancer are second in frequency only to those resulting from cardiovascular disease. Most cancers are newly diagnosed in individuals over the age of 65 and in this age group cancer death strikes most frequently as well. Thus, advancing age and the risk for the development of (the majority of) cancers are directly linked. The uncontrolled proliferation of cancer cells is caused by a failure of the normal process of regulated cell death. With the failure of that regulatory system, the cells are left with an opportunity to divide and multiply without restraint, ultimately derailing the entire organism. This breakdown in achieving controlled and programmed cell death is caused by mutations in a person's DNA. Such mutations are not unusual, but under normal circumstances they are continually corrected before they have harmful consequences. It is when mutations escape repair that the cell that harbors them may begin to proliferate out of control, thus creating numerous cancerous daughter cells. The body typically responds to this with an increased assassination of the new cells, but when further mutations in the cancer cells accumulate, they are oftentimes able to override this compensatory mechanism. When that happens, tumors grow and may spread.

Cell division is controlled by a variety of genes, including oncogenes and tumor suppressor genes. Tumor suppressor genes can be compared to the brakes in a car, because they ensure that cells do not proliferate out of control. When tumor suppressor genes are defective, they are unable to slow cell proliferation and then precancerous cells can develop into true cancer tissue. Oncogenes play an important role in the normal growth and development of the body, and in the cell proliferation that normally maintains the healthy state of tissues. They regulate cell division, but also participate in the process of cell death. In contrast to what their name suggests, these genes do not normally cause cancer, but instead are involved in cancer when their function is disturbed: derailed oncogenes can be likened to an accelerator that is not countered enough, or not opposed at all. Consequently, cell division is no longer properly paired with cell death.

Just like the cellular senescence that results in the aging of our body, cancer is therefore the result of an imbalanced process. If the balance could be restored, then, at least theoretically, cancer could be stopped in its tracks. Achieving this seemingly simple concept, unfortunately, is not that straightforward. Chemotherapeutic drugs that target specific mutations that have been found in the DNA of cancerous tissue often work for a time, after which the cancer acquires additional mutations. It thereby wins out over the drug and can continue to proliferate. These mutations do not emerge in all of a cancer's cells, but only in subsets. Thus, tumors can be quite different between cancers and can also differ internally within a given cancer. This complicates effective, lasting cancer treatment.

As already mentioned, inflammation promotes unfavorable cellular environments. Chronic inflammation is not only a hallmark of age-related diseases but is common in cancer as well. Inflammation-related molecules can promote cell division, tumor invasion into surrounding tissues and blood vessels, and a tumor's tendency to spread around the body. In addition, it appears that systemic inflammation can encourage tumor cells that have traveled to distant sites to settle there and to become metastases. As we age, changes in our immune system may also facilitate the development of cancer, because the immune system becomes less adept at preventing the proliferation of cancer cells. In other words, the aging of the body's cells extends to those in our immune system and this "immune senescence" may enable an escape of cancer cells that would have been caught and destroyed more readily at a younger age. In elderly patients, the immune system becomes less efficient because of a decrease in the number of special cells that can recognize immune system triggers. Such triggers are certain antigens, which are substances that are targeted by the antibodies on our immune cells. Important immune cells are gradually depleted over time, causing a partial but potentially dangerous immuno-deficiency in which cancer has a chance to thrive. So, cellular senescence is a two-headed snake that leads to aging and, as a result of the effects thereof, can lead to the development of cancer.

Much remains to be learned about biological aging, especially because it is a process that involves many steps and encompasses changes at the sub-cellular, cellular, tissue, organ systems, and organism level. It is clear, though, that aging is influenced and directed by a variety of events outside as well as inside the body. There are several hypotheses describing how aging works, each with its own evidence that usually reflects the scientific progress made in one of the areas that is of importance to the aging process. This evidence is not easily integrated and may at first glance appear to involve quite distinct pathways. Science has yet to uncover how all of these models play their roles in the complex phenomenon of aging, but we can look at some major areas of interest that focus on DNA damage, telomeres, and oxidative stress.

Although internal factors contribute to aging in a major way, externally inflicted damage to the genetic material in our cells by exposure to chemicals or radiation is closely intertwined with aging as well. When damage to the DNA in the cell nucleus occurs, it can damage the encoded information that is the blueprint that regulates all body functions. The role of DNA damage in aging has been demonstrated in laboratory experiments with animals in which DNA repair mechanisms were altered, causing early aging. In rare human genetic diseases associated with severe clinical symptoms that resemble premature aging, underlying abnormalities in DNA repair pathways have been identified as a cause. The question remains how DNA damage arises, and also why, given the powerful regenerative capabilities of our bodies, this damage is not effectively repaired but rather accumulated as we age. DNA damage can be induced by many chemicals as well as by ultraviolet radiation from the Sun that penetrates the skin, and by radiation from the radioactivity that is present naturally in our environment or that is introduced by a nuclear disaster. DNA can also be damaged by viruses and other biological agents. Importantly, however, there are multiple processes within our bodies themselves that can contribute to what has been called "genetic stress". These negative effects result from our internal exposure to chemicals that take part in a range of reactions. These include, for example, the reactive oxygen ions that are generated by our metabolism, a process we clearly cannot avoid if we are to remain alive. The free radicals that are associated with these reactive oxygen species can be minimized through scavenging by antioxidants. However, when these are insufficiently available, the oxygen radicals can induce changes to the sequence and structure of our DNA. Once DNA damage has occurred, cell signaling pathways may be triggered that contribute to repair, or that effect the removal of the damaged cells when repair cannot be accomplished. At that point, tumor suppressor genes may be activated to apply the brakes: they intervene by preventing cell division, leaving the abnormal cells stranded and ready for elimination. Sometimes, however, severe DNA damage can override the checks and balances of a cell, resulting in malignant transformation and thereby in the beginning of cancer.

At the larger scale of the entire chromosome, another type of DNA damage occurs as we grow older. The telomeres, which are the chromosomes' tips or endpoints, may not code for proteins but are nevertheless functional parts of our DNA. They form a complex with telomere-binding proteins and are essential to the preservation of chromosome structure and function. Over time, telomeres gradually erode, because the efficiency of the DNA replication system is limited. As a consequence, these extreme regions of our chromosomes are not replicated reliably during cell division. The enzyme telomerase provides some telomere repair but this does not prevent the accumulation of regional damage as we age, and cannot prevent the eventual shortening of the telomeres. Telomere shortening seems to cause cellular senescence once substantial shortening has taken place. It is also possible, however, to have an inappropriate reactivation of the telomerase enzyme. In that case, the genome becomes unstable, tumor suppressor checkpoints do not receive their signals, and tumor development may be initiated. Although telomere length is clearly involved in the aging of our cells, it is not a reliable marker of age. Just like chronological age, which we use to describe how old we are, telomere length by itself is not sufficient to accurately determine the life expectancy of an individual.

Oxidizing free radicals are generally closely balanced by antioxidant enzymes that protect our cells, for example by transforming the radicals into less reactive molecules. Based on that possibility, another hypothesis of aging was developed that involves DNA damage, especially in the area of the cell where reactive oxygen species are being created: the mitochondria, small organelles in the cell that carry their own circular DNA, separate from the DNA in the cell nucleus.

Mitochondria not only control cell metabolism and produce energy for our cells, but are also critically important for growth, differentiation, and programmed cell death, among other functions. The number of mitochondria varies widely between cells and is influenced by the energy demands of individual tissues. Typically, mitochondria have several copies of their own circular genome, which encodes 37 genes that are important in the respiratory chain of the cell. They are very dynamic, pliable organelles that undergo merging and splitting, depending on the energy requirements of the cells and on the tissue type in which they reside.

Mitochondria, inherited only from the mother, are the energy generators within our cells. As such, they are vital to life, and damage of the mitochondrial DNA can interfere with our energy supplies. Yet, they are not equipped with the DNA repair mechanisms that exist for the chromosomes in the nuclei of cells. Thus, DNA damage accumulates over time in the form of mutations that are caused by copying errors in cell division or by chemical or radiation exposure of these small powerhouses.

When someone has inherited one or more mutations to their mitochondrial DNA, serious disease can result that may affect the function of one or more tissues and organ systems, including the heartbeat, muscle contraction, hearing, and vision. Mutations in our mitochondria do not contribute to disease only when they are inherited, however: they are also acquired as we advance in age. Interestingly, as we age, the dynamic splitting and fusion of the mitochondria diminishes and their numbers are reduced while their sizes increase. This is thought to be a compensatory mechanism that maintains the total volume taken up by mitochondria, but may not, in fact, preserve vitality. The extent to which mitochondrial DNA damage directly causes aging is as yet unclear, but the view that mitochondrial dysfunction contributes to aging and age-related conditions is firmly established and widely accepted.

To gain a full understanding of how and why we age, aspects of the various hypotheses of aging (including DNA damage, the role of telomeres, and oxidative stress) are being studied. Some researchers prefer to focus on studies of longevity or investigate diseases that highlight certain aspects of aging. Examples of the latter are the rare progeria syndromes, which are a diverse group of conditions that have in common an association with an accelerated although incomplete aging process. Each of these conditions is characterized by a specific set of symptoms, some of which are like changes seen in elderly individuals. Patients with these diseases have premature aging of a subset of their tissues. They may still be children yet appear old, and may look emaciated or bent over because of osteoporosis, muscle weakness, or a curvature in the spine. They may appear frail, prematurely bald, have strokes, or experience early loss of vision and hearing, and they also have a reduced life span. These patients have severe conditions that are caused by mutations in genes whose normal function it is to oppose the effects of aging. These mutations include those that cause the DNA and telomere repair processes to falter.

The effects of aging are not limited to personal or academic interest, but have a major societal impact. The number of persons over the age of 65 will drastically increase over the coming decades, changing the makeup of our society. With the increase in the population of older people we can anticipate a considerable increase in the number of age-related problems, unless, of course, we find a way to lengthen people's health span. Apart from much-needed education about the benefits of a healthy lifestyle so that people have the knowledge to make better choices, we need to make the tools available to help people reasonably accomplish this. Biomedical research is also expected to result in direct benefit by being translated into practical health improvements. In particular, there is hope that injured tissues and diseased or abnormally functioning cells can be replaced by embryonic stem cells or by adult stem cells. These two types of stem cells have their advantages and disadvantages, but both can differentiate into required cell types and both have great capacity for regeneration and continued renewal. The administration of antioxidants and other supplements or medications may also mitigate the effects of aging.

In the end, there is no permanent way around aging. Knowing just how much goes on in our bodies to maintain it in a healthy functional state over several decades, it is simply amazing how well the body functions. Much in our body is replaced, recycled, and in flux over our lifetime. Billions of cells die each day and roughly the same number are created in cycles that continue until we reach the final limit of this capacity and our bodies, with our death, are entirely returned to the Earth.

Key Points: Chapter 3

- As we age, the capacity of our cells to replace themselves and regenerate decreases. An excess of cell death over cell renewal leads to physical decline.

- The continual replacement of cells by new generations enables repair of damage, but over time the effects of internal and external influences (including some lifestyle choices that we make) lead to increasing problems for the successful functioning of the cell.

- Advanced age increases the probability of getting a variety of diseases. When deterioration persistently outpaces repair, cellular loss impairs body function and ultimately causes organ failure. Failure to repair can also derail into unrestrained cell proliferation. This is the hallmark of cancer.

- Our DNA contains instructions for cells to speed up or slow down the rate at which they divide, normally leading to growth in development and a healthy stability in adulthood. But when chemicals, radiation, or other influences damage DNA, checks and balances may be upset, leading, for example, to cancer.

- **An imbalance between cell death and cell creation leads to growth and development, but also to disease, aging, and death.**

Tell me what you eat, and I shall tell you what you are.

JEAN-ANTHELME BRILLAT-SAVARIN (1755–1826),
in *The physiology of taste* (1825)

As cellular units are built up from the elements of the food
accessible to them, and as the nature of the food influences
their state of well-being, [. . .] animal and vegetable bodies,
forming collective groups of cells, are as dependent upon the
elements given them in food as any single-celled representative
of individualism.

ALEXANDER LOCKHART GILLESPIE (1865–1904),
in *The natural history of digestion* (1898)

Food for Thought

We rarely stop to think that everything that we are composed of changes continuously in major and minor ways, day by day, hour by hour, and second by second. Most of the biochemical and physical processes in our body occur in tiny fractions of a second, far faster than the life cycles of the cells within which they happen. In this chapter we take a look at the chemical substances that are the basic building blocks out of which our bodies are composed and we consider where they come from, how they flow through us, and how they are made or used by our cells.

Whereas our cells, depending on their function, may stay with us for a time span from days up to an entire lifetime, our body at any given moment contains much that has entered it just a short while before. Most of the vegetables, fruits, and meats that we consume are in the form of clusters of cells. Those, and anything else we eat, in the end consist of molecules: small particles that are composed of a variety of atoms linked together by shared electrons. Atoms, in turn, constitute the smallest unit of a chemical element, which in essence is matter that cannot be split further by chemical methods.

Water is the most common chemical in our bodies. Each water molecule is composed of two hydrogen atoms and one oxygen atom, linked together by a polar covalent bond, one of the many possible kinds of chemical connections in which atoms share and exchange their electrons. This particular arrangement, though, makes water a remarkably good solvent for salts and sugars, proteins and DNA, alcohol, and even gases such as oxygen and carbon dioxide. Its solutions can be acidic or alkaline. Moreover, water is a low-viscosity liquid at the temperatures found over most of the Earth's surface. As a result of its chemical and physical properties, water is a medium in which other chemicals can be mixed and brought into contact with each other in a large variety of chemical reactions. Water is not a good solvent for fats and oils, but life has found solutions to work around that problem, including wrapping molecules of fats and oils inside other molecules that are soluble in water, or by attaching strings of atoms to fat molecules that make the composite soluble in water.

About two-thirds of the water in our body is contained within cells, whereas one-third exists between cells. Overall, water amounts to approximately 60% of the weight of the human body, depending somewhat on a person's age and weight. Thus, it truly is the juice of life. The importance of water to the human body is illustrated by the fact that whereas a person can be without food for weeks at a time (obviously getting hungry and starving slowly), lack of water leads to rapid deterioration and can cause death in a matter of days.

Most chemical reactions in our cells occur in an aqueous environment, which is to say that they occur within the liquids of the cell. These liquids in cells consist largely of water that is enriched with a variety of chemicals that keep healthy cells from swelling or shrinking too much when our level of hydration changes. These additives to the water inside cells enable the body to use the liquids and nutrients that it takes in, but it is not forced to absorb them immediately and completely. Instead, the flow of liquids in our body is controlled by cell walls and is strictly regulated by a variety of biological processes.

Apart from the hydrogen and oxygen that constitute water and that are part of other molecules, the elements carbon and nitrogen are also abundant in our bodies. All together, these four elements add up to more than 96% of our body weight. It seems almost impossible and is really quite humbling that four simple elements are able to construct most of what we humans are made of, and that these four enable an astounding assortment of physical functions and behaviors.

The chemical makeup of all life on Earth is dominated by the same four elements. There is an entire field of chemistry that is named after this basic similarity: organic chemistry, which is the chemistry of life. Organic chemicals always involve carbon, which is the foundation of the living organism. Although more than 20 elements help shape organic molecules (we turn to that in Chapter 6), hydrogen, oxygen, and nitrogen are the most common elements found in organic molecules. Approximately 3% of our weight is made up of nitrogen. This element is the least abundant of the set of four, but it is present in almost all of the carbon-containing complex molecules in our bodies. In part because there is so much water in our bodies, oxygen dominates by weight, accounting for 65% of the total. Although there is even more hydrogen by number, these atoms are so light that they amount to no more than 10%

of the body's weight. Carbon, lastly, accounting for 18% of our weight, is the key element of the solid compounds in our body.

When we consume food and water, we take in all these elements, but they do not come as individual atoms. In fact, we could not use many of them as nourishment if they were mixtures of individual atoms. Instead, what our bodies mostly need are mixtures of organic molecules, which are complex groups of atoms (carbon and various other elements) bound together. Rather than focusing on the elements here, we discuss the most important types of molecules that our bodies need. Organic molecules can be distinguished by their overall size as well as by their various chemical groups that give the molecules specific characteristics that enable them to carry out one function or another within the cells. Some organic molecules are utilized as building blocks for macromolecules that are assembled from simpler chains, whereas others just provide energy to our cells. The ultimate distribution of molecules in the cells all over the body depends on the predominant need at any given time.

Because of their diversity in shape and chemical properties, carbon-based molecules have a host of applications in different cellular environments of the body. These organic molecules are grouped into four main categories: carbohydrates, lipids, proteins, and nucleic acids. Their functions include carrying and storing energy, forming the building blocks of the cells, regulating the body's chemistry, and storing information from which cell structure and function are derived. With so many molecules involved in the process, and with many performing multiple tasks or acting as parts of others, the separation of function is not nearly as clearly mapped out as their separation by structure and content, which is what the grouping by main category does.

First, we look at the carbohydrates. This term literally means "watered carbons", a name that reflects the fact that they generally contain two hydrogen atoms for every oxygen atom, which is the ratio in which they occur in water. The simplest of these carbohydrates are known as sugars, which are diverse but structurally similar molecules that contain carbon, hydrogen, and oxygen in particular configurations. Glucose is a sugar that provides energy to our cells. Other types of carbohydrates, such as glycogen, are used to store energy. Carbohydrates represent a fuel that can be mobilized swiftly when needed, but they also have a range of additional purposes in the body, particularly when they are combined with other molecules. For example, different sugars on the cell surface help determine our blood groups. Sugars are also a critical component of the very backbone of our DNA.

The second category is that of the lipids. Also known as fatty acids, lipids contain a special chemical group that has affinity with water, even though the rest of the molecule shuns it. Lipid molecules vary in the number of carbon atoms in the chain as well as in the number of chemical bonds that hold the molecule together. They can have short chains or long chains, and they can be unsaturated or saturated, which reflects how tightly the molecules are packed. Lipids are very important for the fabrication of semi-permeable membranes, which largely isolate particular components within the cells, as well as for the creation of functional structures within cells where most of the actions of cellular metabolism take place. The membranes are made of a lipid double layer that repels water at its center and has affinity for water at the surfaces that are exposed. These membranes cannot be crossed by most hydrophilic (or "water-loving") molecules because of their water-repellent center. Structures inside the cells that are cordoned off by such membranes include the Golgi organelle, which cleans up the waste inside the cell, the smooth endoplasmic reticulum, which transports nutrients and fluids, the rough endoplasmic reticulum, which is the site of protein synthesis, and the mitochondria, which are the cell's industrious engines, where energy is generated for the cell to use.

Fatty acids are an important source of energy, just as glucose is, and also help store energy in our fat. Lipids can be stored inside cells and then be broken down into usable compounds when the body needs these reserves. When a person is overweight, it reflects a shifted balance of food intake and physical activity or energy usage. The excess food is converted into fat, which is accumulated in the adipose cells, which specialize in storing fat. They grow in size when too much energy is taken into the body.

The third category of organic molecules is that of the proteins. Proteins (together with nucleic acids, to which we return below) are the most abundant materials in our cells after water. The protein molecules in our bodies are assembled from 20 different amino acids. These amino acids all have two chemical groups in common, each on one end of the molecule, with a variable number of carbons and chemical groups on side chains in between. Through their incredible diversity of shapes and interactions, proteins enable an extraordinary array of functions within our cells. In part, that function depends on the amino acid sequence and chemical characteristics of the protein molecule, but it is also based on the functional domains and the three-dimensional, folded shape of the molecule. Proteins can form remarkable complexes, with moving subunits. They can take on highly specialized properties and can interact with multiple other large molecules. Proteins, encoded by our genes, not only are utilized as the basic structural building material for much of our body, but they also manifest as hormones, as antibodies, and as transport molecules. Important protein roles include cell signaling, growth, development, wound healing, regulation of the majority of chemical processes in the body, and energy generation.

Proteins can take on a structural role, for example by linking together to shape what is called the extracellular matrix that glues our tissues together. Or they can work as mechanical drivers that help our muscles contract when we move. Proteins can also act as enzymes. Enzymes are an important class of proteins that significantly speed up chemical reactions in our cells without being altered themselves by those reactions. One example of a reaction in which an enzyme is critical is when we process a pharmaceutical drug in the liver to get to its active, useful components. This drug may have been given to a patient in order to inhibit blood platelets from aggregating, thus effectively preventing blood clotting after a surgical procedure for heart disease. Without the essential enzyme, the drug would not work. This can be demonstrated in a laboratory experiment but can also be observed naturally in some patients who happen to have a change in their DNA sequence for that particular protein, resulting in a slightly changed enzyme structure. Such a modified enzyme does not work as efficiently as the unaltered enzyme that is present in people who do not carry the variant. When the DNA of a patient is analyzed by, typically, a pathologist and the result shows that there is a change that makes the enzyme ineffective, then another drug that can be processed by the body can be prescribed instead, to minimize the risk of blood clotting for that particular patient.

The fourth category of organic chemicals is that of the nucleic acids. These are the modules that build our DNA (deoxyribonucleic acid), which forms the instruction code or template for proteins; we have in total a few hundred grams of DNA in our bodies, roughly the weight of an apple or two. RNA (ribonucleic acid), which essentially comprises the action messages transcribed from the DNA code, is also a nucleic acid. The building blocks of nucleic acids are called nucleotides. Our DNA has a code that is based on a "four-letter alphabet": A for adenine,

C for cytosine, G for guanine, and T for thymine. The order in which these four "letters" occur dictates the information by which cells are built and by which they function. This is known as the DNA sequence, which can be deciphered letter by letter.

Genetic diseases are commonly caused by a single "typo" in the sequence, when just one nucleotide was somehow replaced by another. Such small molecular exchanges constitute errors within the DNA and alter the transcribed RNA message that tells the cell how to build a particular protein. This can deform the protein structure, and thereby alter or inhibit its function, thus causing disease. An example of this is a change in the DNA sequence that encodes a protein that is responsible for the processing of cholesterol-rich fats. As a result of just one single nucleotide error, the instructions for building that protein can become much shorter than they should be, thereby ultimately thwarting the normal processing of these fats. As a result, the person with that simple mutation now is a patient with a hereditary disease and high levels of cholesterol, which are in turn associated with an elevated risk of heart problems. In such patients, the consequences of high cholesterol levels are likely to occur at a much lower age than that when it is usually seen in patients who have high cholesterol due to other causes.

Although protein function can often be predicted from the DNA sequence, as in this example, it is not always straightforward to foresee the effect of changes in the genetic code. This can pose a diagnostic conundrum, because the extent of the effect can differ according to which nucleotide was replaced and what other nucleotide took its place: some changes turn out to be clinically innocent, whereas others can cause severe disease.

The nucleic acids in the cell nucleus form long strands that, taken together, comprise our genetic code. This genetic blueprint is the basis of growth when we develop into adults, and of our ongoing cell and tissue renewal throughout life. It contains the information for the synthesis of proteins, which contribute to the structure of our tissues and perform biological functions in our cells. DNA is replicated when needed during cell division. Errors in the replication process are usually corrected with high efficiency, so that the copies are reliable and enable the continued proper functioning of the human body. Although our genetic code is quite detailed, the actual weight that nucleic acids add to a cell is virtually negligible. When we look at the length, instead of the weight, of the totality of our nucleic acid strands, we realize how amazingly efficiently information is packaged into DNA: very tightly bound to take up a minimum of space, each of our cells has a total of approximately two yards (or two meters) of DNA in its nucleus alone. Additional nucleic acids can be found outside the nucleus, in the cytoplasm of the cell. One such place is in our mitochondria, which are critical to our cellular energy supplies. Leaving the nucleic acids outside the nucleus aside for now, however, fitting in two yards of stringed nucleic acids truly is an example of superior packaging, given that the average human cell is only about 50 micrometers (50 millionths of a meter) across. For the body as a whole, the numbers become even more astonishing. Our super-coiled DNA, if uncoiled and stretched out in a string, could wind around the world over two million times, or loop to the Moon and back 130,000 times, and even stretch to the Sun and back roughly 333 times!

Now that we know the four major categories of organic molecules (carbohydrates, lipids, proteins, and nucleic acids), let us a look at the processes in which they are involved that make the body function. Cellular processes often require enzymes to be optimally efficient. Enzymes (generally proteins) and the substances on which they act (called substrates) bind to each other at incredible speeds within the cell. This can happen because the substrates, which are much smaller than the enzymes, move very rapidly through the cytoplasm that fills the cells. Once such a substrate molecule encounters the right kind of enzyme, the two can associate and react almost instantaneously. Thus, individual enzymes are like turbo engines that catapult specific reactions forward. When multiple enzymes join forces in metabolic pathways that encompass a series of chemical and physical processes, then, instead of just single reactions, entire chains of reactions are enabled or immensely accelerated. This synergy is what allows metabolism to occur in a way that greatly speeds up those reactions, in a vast network of pathways that are needed to sustain life. This is not just relevant inside a given cell, but also aids in the transfer of small molecules through pores in the cell wall, which enables the exchange of materials and establishes communication between cells.

Metabolism reflects the dynamic, ever-changing state of our cells. Through the network of metabolic pathways, energy is released to the body when needed, and surplus energy is stored for later use. When a surplus of energy is available it can be used to build and store sugars, lipids, and proteins. When energy or building materials are needed, these stores are tapped to help break down larger molecules into their smaller components such as simple sugars, fatty acids, amino acids, and, ultimately, waste products. Our individual cells contain sophisticated chemical machinery that works diligently to fuel body functions. In order for metabolism to function properly, however, the body needs to be appropriately nourished, must be able to handle its intake, and should be efficient at excreting waste products. The combined effects of metabolic processes within each individual cell result in a large-scale operation that becomes obvious at the level of our entire body by the elimination of substances such as urine and stool, and by breathing.

The diverse manifestations of metabolism include the means by which the cell takes in liquids and those through which the cell encloses solid materials. To release substances, cells typically use the Golgi organelle, which is specialized in the packaging, sorting, and transporting of cellular stuff. If we want to understand what happens to certain molecules in our cells, we need a way to tag and trace them. This can be done in research laboratories by attaching a tracer to the molecules of interest. This can be a fluorescent segment or a radioactive atom; the choice of tracer depends on the scientific question posed and the type of studies performed. Such tagged molecules either glow or trigger instruments sensitive to radioactivity, thus enabling both instant detection of the molecules under study and the tracking of such molecules over time and in different tissue compartments. It would be too difficult to follow all molecules in a cell at the same time, because each small cell is a world of its own. Determining exactly, then, how all molecules inside a cell are made, used, exchanged, and destroyed in a single view is impossible for the time being, but many aspects of the overall metabolic process can be understood by focusing on smaller steps and by following specific classes of molecules.

For all its metabolic functions to be possible, our body needs a continuous supply of energy, derived from the nutrients provided to the body. These, in turn, are sustained by large-scale processes that take place in our environment and that reach far beyond the Earth: our capacity to live is entirely beholden to photosynthesis and global respiration, neither one of which would be possible if our planet did not have the benefit of the Sun's supporting light.

Photosynthesis is the process that harnesses the energy in sunlight in the form of photons. This energy is taken up by chlorophyll-containing proteins in plants, and enables the conversion of carbon dioxide and water into sugars and other organic molecules that plants need for growth. Animals, including humans, subsequently use those substances from plants as food, either by directly consuming the plant materials or by consuming animals which did so. Our Sun is not only at the heart of our nutrition, but also gives rise to the oxygen released from plants as a byproduct of their photosynthesis. Although a waste product of plant life, oxygen is what makes it possible for us to breathe and live by digesting the plants that made it. Our respiration through the lungs permits the oxygen from the atmosphere to enter our blood stream, by which it is transported to all of our tissues. Within the tissues, we use it to oxidize, or "burn", organic molecules. Most of the molecules used for this process are sugars, especially glucose. The reaction produces chemical energy for our cells, and in the process releases water and carbon, which are removed from our bodies as waste chemicals.

Energy generation in any animal requires oxygen, but obviously we cannot live on air alone. We must eat in order to obtain the nutrition that can be converted into the energy needed for all of the functions of our body. Although we rarely think about it in this way when we enjoy a meal, it is that same meal that is employed to maintain the biological tasks of our cells and that allows us to survive when energy resources are expended, when cells need to be restored, healed, or replaced. When food arrives in the digestive tract, it is first broken down from large food particles into smaller ones by mechanical and chemical processes in the mouth, in the stomach, and in the first part of the gut, known as the small intestine. This digestive process is aided by the acid in the stomach and by a multitude of enzymes, by secretions of the liver and the pancreas, and by hormones.

Subsequent to their first crunching, the nutrients must be absorbed from the gut into the body. This transfer occurs largely in the small intestine, which is relatively narrow compared to the large intestine, which forms the last part of the gut. Despite its name, however, the small intestine is not small at all: it has a total length of about eighteen feet (six meters) and an enormous surface area—approximately 2700 square feet (250 square meters), which would easily cover the floor space of a sizable house. This surface area fits into our bodies through another example of very efficient packaging: because it has a highly folded, corrugated lining, the small intestine packs a lot of surface into a tight space.

The physical barrier provided by the intestinal surface is structured in a way that permits absorption of nutrients. The surface area of the small intestine is colonized by thousands of species of microorganisms that are advantageous to the normal functioning of the digestive process and, therefore, to our health. This reflects a fully symbiotic relationship, in which we provide food and a livable environment to those bacteria, while they perform essential functions to pre-process our food into usable forms. When we eat, we do not merely provide for ourselves, but also for a host of other life forms that live within us. Apart from aiding in digestion, the beneficial bacteria in our gut protect us from over-colonization by harmful competitors. If these beneficial bacteria were to be allowed to move through the intestinal lining though, they would most certainly be detrimental. To prevent that from happening, the membranes in our gut that provide the surface for absorption are well protected by the immune system.

In order to understand how our bodies utilize the food we eat, it is helpful to briefly review what we consume. Our foods comprise different mixtures of the four major categories of organic molecules and thus mainly consist of carbohydrates, proteins, lipids, and nucleic acids. Much smaller amounts of vitamins and minerals are also ingested. Carbohydrates can be found in a variety of foods, including whole grains, beans, and other legumes, such as peas, soy, and peanuts, as well as in fruits and vegetables, including starchy vegetables such as sweet potatoes. Complex carbohydrates provide excellent nutrition to our bodies, and have only a fraction of the calories contained in fat, while many that are non-digestible promote healthy digestion as fiber. These attributes are also helpful to maintaining a healthy weight.

The complex carbohydrates differ from the simple sugars such as glucose that are often found in processed foods such as white flour and sweets. These processed foods have been depleted of the most valuable nutrients and have much less fiber. They have a high glycemic index, which means that the consumption of such foods results in rapid glucose absorption, causing the spiking of blood sugar levels. That can be taxing on the body and typically requires a boost of insulin from the pancreas to move the surplus blood sugar into the cells, to prevent it from being harmful. This glucose buildup is the basis of illness in diabetics, who are either unable to produce insulin or whose metabolism is not responsive enough to it. A state of high blood sugar levels can be noticed in the body by healthy people too: it brings about a feeling of sluggishness and fatigue after eating, rather than an energized feeling when the body receives healthy "fuel". All types of carbohydrates are eventually digested into small sugars by enzymes in the intestinal wall, with the help of local, beneficial bacteria.

Lipids are also an energy source for the body. They are typically consumed in the form of triglycerides and are energy dense. Found in fatty foods, they provide nine calories per gram, compared with only four for carbohydrates and proteins. Even as it is recommended to keep fat intake low overall, it is preferable to select unsaturated fats derived from plant-based foods such as olive oil, over the saturated fats found in meat and dairy products. The latter can contribute to elevated levels of cholesterol and are thereby a major contributor to heart disease. Lipids are processed in the digestive tract through the use of bile and enzymes that help with their breakdown into fatty acids and simple chemicals, which are then absorbed through the intestinal wall. In our body, fatty acids are used for the generation of membranes and other tissue components, for the construction of biologically active substances, and directly for energy. Fat can be stored in the cells of the adipose tissue, of which we each have several dozen billion. When one gains weight as a consequence of higher energy intake than energy use, the excess fat is stored there. Unused sugars are also converted into fat, which is then stored in our adipose cells until it needs to be converted into energy.

The next category of major organic compounds that the body needs as nutrition is that of the proteins. These are found in nuts, vegetables, beans, soy, seeds, eggs, dairy products, and meats. They are digested into smaller pieces called peptides and further processed into their amino acid building blocks, which are absorbed into the blood stream and provided to our cells, where they are used to build protein chains as instructed by our DNA. This process can ramp up or down as needed. There are nine amino acids that cannot be created within our bodies from smaller components, and these are called the essential amino acids. They are supplied to the cells only through our food. Proteins are critical nutrients, but a healthy diet routinely provides them in ample amounts. Interestingly, proteins can be converted into other molecules and used for energy, but they cannot be stored for a longer period of time themselves. The perception that a high-protein diet is unequivocally healthy is, therefore, misguided. Rather, excess protein in a diet can be a taxing burden on the purifying function of the kidneys and can contribute to adverse health effects such as kidney stones.

Last but by no means least on the nutrients list are the vitamins and minerals that contribute to our health. Rather than being used as a direct source of energy, they assist in the processes of energy generation. The difference between vitamins and minerals is that vitamins are organic in nature, whereas minerals are inorganic elements that are taken up by plants and then ingested by humans through consumption of the plants (unless directly consumed as pure minerals added to our food, such as salt). Minerals have a wide variety of functions, ranging from the maintenance of an electrolyte balance, support of muscle contraction and food digestion, and transport of oxygen, to growth, development, and bone mineralization.

Vitamins are organic compounds that the body cannot create enough of by itself, but of which it needs adequate amounts to enable essential chemical processes. Vitamins help build proteins either directly or as stimulants to enzymes. Among the functions of vitamins are contributions to energy metabolism and to healthy bones, teeth, nails, and skin. They also support our nervous system, blood cells, blood clotting, and DNA synthesis. Twelve vitamins are known to be essential to humans, many of them having multiple functions within the human body. We must maintain the right amounts of all of these vitamins in our bodies by consuming organisms that do have chemical pathways to create them, including bacteria, plants, and fungi. One other group of chemicals is generally added to this list of 12, known under the name of vitamin D. Humans can make their own vitamin D if exposed to enough sunlight. It is the only chemical in our body made by direct exposure to the ultraviolet component of sunlight. Vitamin D can be found in food, too, though, especially fish, eggs, mushrooms, and liver. For dogs and cats, however, it really is a vitamin: they cannot synthesize it themselves.

Some vitamins and minerals exist together. Cobalt, for example, is a trace mineral that is a constituent of the B12 group of vitamins. No animal can incorporate that mineral into the synthesis of those vitamins, and therefore all need to ingest it from the kingdoms of bacteria and archaea (including in many cases from those living in their guts). Cobalt cannot be utilized by the body's chemistry if it is ingested separately.

Our need to maintain the diversity of vitamins and minerals in our bodies can be met only by a diverse diet. Limiting diets to particular food groups can result in diseases. Entirely inadvertently, such dietary restrictions led to the discovery of vitamins and their roles in the body, although at substantial cost. Among the better-known stories of vitamin deficiencies is that of vitamin C. Removing fresh plant food as well as raw animal flesh from a diet leads to the typical vitamin-C deficiency disease called scurvy. Scurvy was a common seaman's disease from the moment long sea voyages were undertaken. In the late fifteenth and throughout the sixteenth and seventeenth centuries, ships could lose the majority of their crews to scurvy. In areas of the world where fruits and vegetables are scarce, such as the Arctic, the consumption of raw meats traditionally provided for people's need for vitamin C. This is because just about all animals, apparently excepting only humans and guinea pigs, can create their own vitamin C. This illustrates that what is an essential food ingredient to one species is not necessarily that for another: lists of vitamins are specific to a species. Moreover, even though human vitamins may be created in different organisms, the detailed molecular structure of that vitamin may depend on the organism that creates it: a vitamin is defined by its function, and multiple molecular shapes can meet the requirements for that function. For example, the generic description "vitamin A" has at least six known distinct molecular shapes synthesized by different plants and animals that we consume, and can be used almost, although not fully, interchangeably within the network of human biochemical pathways.

Our bodies need energy all the time, but obviously we are not continuously eating, so the body has to deal with the time between meals and, periodically, with extended periods of fasting. Storage of foods for later conversion to energy is therefore a priority. When there is more than we need for immediate use, cells such as those in the liver and muscles can store carbohydrates in the form of glycogen, while adipose cells can store fats in the form of compact lipid droplets. Our first source of energy is glycogen, but fats can be used when the energy needs surpass those of our glycogen reserves, and when the need for internal energy generation from stored sources, rather than from externally provided nutrition, continues over a longer period of time. Protein cannot be stored in the body for an extended period but can be digested into amino acids, which can be converted into sugars and stored with the help of enzymes, which feature so prominently in almost all of our cellular functions. Excess protein is typically, however, processed in the liver and then excreted through the kidneys. When there is a sustained shortage of energy and the sources of both glycogen and fat become depleted, the body resorts to metabolizing the proteins from our muscle mass.

The overview provided here is far too short to reflect the actual complexity of the numerous pathways that convert nutrition to energy. Nevertheless, it offers an impression of the chemical machinery constantly at work to keep us going in times of abundance and in times of scarcity. It has been suggested that our likes for fatty and sugary foods are directly linked to the ancient need to take advantage of times of plenty, when food was readily available. In times of shortage, the body would then have the reserves to sustain itself. Today, however, at least in the developed world, most people live in an environment in which calories are abundantly available, causing obesity to be one of the most concerning health risks in these societies. Now, more than ever, it is important to distinguish ready access to calories from a true wealth of nutrients. Health is determined not only by the amount of calories we consume, but much more importantly by the quality and diversity of the nutrients we make available to our body, so that these can be used as fuel for its many tasks. One of these tasks is to stabilize the energy balance for our system in different circumstances, in situations of health as well as illness, in times of ready access to food and in times of need. The flexibility and resilience of our system are quite remarkable: because of the many regulatory processes that use the energy available at any given time, we can maintain a reasonably stable equilibrium that is relatively independent of the whims of our environment.

The food that we consume, the energy and building materials that it provides us with, and our overall health status are closely related. If we eat too much or if we consume unhealthy foods for long periods, we likely curtail our life span. The detrimental effects of caloric overload and obesity are evident from the associated high frequency of coronary artery disease, diabetes, and even some forms of cancer. Investigations in organisms as different as yeasts, worms, flies, and primates have demonstrated that a healthy overall lifestyle combined with modest calorie intake may, in fact, increase life expectancy, while it delays or even avoids the onset of such diseases as well as other age-related illnesses. It may also delay the loss of structure and function that is associated with aging overall, and it may prevent frailty. In addition, moderate calorie intake, provided that it is with adequate nutrition to stay healthy, is one example of the mild but beneficial incentives to our body that leads to an adaptive response and makes our cells more resistant to bigger environmental pressures. This flexibility improves our chances of longer survival.

The extraction of usable energy from our nutrients is heavily influenced by the mitochondria. These play a central role in the use of oxygen and in the power supply for our cells. Mitochondrial function is affected by both genetically encoded instructions and our body's needs, which depend on the environment in which we find ourselves. If there is an abundance of calories and little physical activity, more harmful components will be generated and there will be more oxidative stress to our cells, more accumulation of mitochondrial mutations, and an increase in the risk of age-related diseases. A healthy lifestyle is better not only because of the benefits of fitness, stable blood sugar, normal blood pressure, and healthy levels of blood lipids, but also because it prevents the generation of substances that are damaging to the mitochondrial DNA in the very energy engines that we rely on for our survival. That survival depends on taking in the right diversity of chemicals to power and build the body; chemicals that are themselves used, modified, and built in the world around us, where some of the energy is available in forms that we, humans, cannot directly consume, but for which we rely on other life forms that can.

Key Points: Chapter 4

- The elements hydrogen, oxygen, carbon, and nitrogen make up to more than 96% of our body weight. The chemical composition of all life on Earth is dominated by these same four elements.

- Carbon is the key component of organic molecules, in which it is combined with about 20 other elements to form the basic building blocks of our bodies.

- There are four main categories of organic molecules: carbohydrates, lipids, proteins, and nucleic acids. They carry and store energy, form building blocks of the cells, regulate the body's chemistry, and help store the information from which cell structure and function are derived.

- A network of metabolic pathways releases energy to the body when needed and stores surplus energy for later use. This chemical machinery constantly keeps us going.

- Our body needs the energy derived from nutrients in our food for all its metabolic functions. Survival depends on taking in the right diversity of nutrients to power and build the body, chemicals that are themselves used, modified, and constructed in the biological world around us.

- **The food that we consume, the energy and building materials that it provides us with, and our overall health status are closely related.**

For when the sun draws up some vapors here, or warms a plant there, it draws these and warms this as if it had nothing else to do.

GALILEO GALILEI (1564–1642),
from *Dialogue concerning
the two chief world systems* (1632)

May not light also, by freely entering the expanded surfaces of leaves and flowers, contribute much to ennobling the principles of vegetables?

STEPHAN HALES (1677–1761),
in *Vegetable statics* (1726)

Basking in Solar Energy

The energy that powers our cells is extracted from the food that we ingest. Within the biosphere—the combined terrestrial regions that sustain life— that energy ultimately derives from plants: we either consume plants directly or we eat animals that consumed plants before they were used as human food. Even most of the energy that powers human society comes from plants or from the animals that ate them, being exploited by us in the form of fossil fuels, for example. Ultimately, all of that energy, including the renewable resources that rely on moving air or water, originates in the gigantic nuclear furnace deep within our neighboring star, the Sun. How is all of that connected?

Energy may be a concept that appears fairly straightforward to grasp intuitively, but this sense soon runs into trouble when we try to understand why energy comes in many different forms. Further, one form of energy can convert into another of an apparently entirely different nature. For instance, energy can exist as the motion of matter but also as the electrical pattern of bonds between atoms; it can take the form of light and can exist in the form of a temperature; it can be taken from electricity or magnetism or gravity; even matter itself can be converted into various forms of energy. The conditions under which one type of energy transforms into another determine how efficient the process is. Seen from the perspective of human biochemistry and human society, the ultimate source of energy for virtually all life on Earth is sunlight, which ultimately comes from nuclear fusion within the Sun. The energy released by the Sun goes through a complex string of transformations before it enables us humans to move, to stay warm, to read, to think about daily life, or to ponder the reasons why the universe exists.

Within the innermost 25% of the Sun by radius (or only about one-sixtieth of the Sun's interior by volume) the density and temperature of the gases of which the Sun is composed are both high enough that the cores (nuclei) of hydrogen atoms frequently collide. Nuclear collisions like that do not happen easily under conditions that we, humans, consider ordinary: at room temperature it is simply impossible for two positively charged atomic nuclei fly into each other fast enough to overcome the electric repulsion between them. In the core of the Sun, however, the temperature is some 54,000 times higher than our comfortable room temperature. Physical laws tell us that at that temperature the gas particles move 230 times faster than they would at room temperature. That is enough for a hydrogen core to have a very slight probability of colliding with another nucleus. For any one hydrogen core that probability is very, very small, but when matter is packed as closely as it is in the center of the Sun, which is over 130,000 times denser than in our atmosphere at sea level, the total number of those nuclear collisions does add up to a rather considerable number: 620 metric tons of hydrogen is fused in nuclear collisions deep within the Sun every single second.

Through a series of such fusion collisions, interspersed with decay into different combinations of components, sets of four hydrogen nuclei fuse into cores of helium atoms. This process has been going on for 4.6 billion years in the Sun and will go on for another 4.5 billion years before all of the hydrogen deep inside the Sun has been used up and converted into helium. The energy released in this process causes the surface of the Sun to continuously emit 63 megawatts per square meter; for an object with a radius of 700,000 km (435,000 miles, or just over 50 Earth diameters), that makes for a truly astronomical total energy output. Although that overall energy flux is enormous, so is the volume within which that energy is being generated. Dividing a large rate of energy generation by a large volume means that every cubic meter within the core of the Sun involved in nuclear fusion generates on average merely 18 watts, which is about the same as an energy-saving household lamp and—gram for gram—only a quarter of the power that a reader of this text, seated in a chair, generates by chemical reactions in the same unit of time. So, it is not the fusion rate per unit of mass, but rather the vast size of the Sun that results in its gigantic energy output.

Nuclear fusion converts one chemical element into another by clumping nuclear material together. This process transforms part of the mass of the particles involved into other forms of energy. The fundamental basis of this conversion was captured by Albert Einstein in a famous equation first published in 1905 that expresses the equivalence of mass (m) and energy (E), connected through the square of the speed of light (c): $E = mc^2$. The full implication of $E = mc^2$ for nuclear reactions in stars became clear only more than three decades after Einstein first published it. Other things had to be discovered first, including the existence of a short-range but strongly attractive nuclear force that binds the nuclear particles together even as their electric charges work to repel each other. It also required the development of the science of quantum mechanics and—in 1932—the discovery of the neutron. Only with all that in place by 1939 could Carl Friedrich von Weizsäcker in Germany and Hans Bethe in the United States work out the most important reaction pathways by which hydrogen fuses into helium deep inside stars.

They worked out that, in a star like the Sun, the fusion of four hydrogen nuclei into one helium nucleus involves a series of nuclear collisions. Together these collisions produce one helium nucleus, two neutrinos, two short-lived positrons and two gamma-ray photons. Neutrinos are remarkable particles that move at very nearly the speed of light and as they travel they manage to ignore just about everything that they fly through. The vast majority of all neutrinos that are generated by nuclear fusion in the solar core fly right through the overlying layers of the Sun, through the near-vacuum of space (not surprisingly), and through entire planets, should they encounter these on their way out of the solar system. Neutrinos fly almost effortlessly through anything they encounter, including the Earth and anything and anybody on the planet. Hundreds of billions of neutrinos pass through a human body each second with very, very few ever getting stuck inside. On rare occasions, one of those neutrinos does interact with the matter that it travels through, however, and very large neutrino detectors, filled with many tons of specially prepared chemicals, may be able to capture a handful of neutrinos in a day. Despite these very low numbers of neutrino interactions with ordinary matter, neutrino detectors proved to be very important in confirming that we do indeed understand the internal structure of the Sun in detail, as well as the processes of nuclear fusion that occur near its center.

Another product of solar hydrogen fusion is the positron. Positrons are the anti-particles of electrons: when a positron encounters an electron, the two immediately transform into a pair of gamma-ray photons. Positrons can exist only briefly, only as long as it takes between its creation and a collision with its anti-particle. Because the Sun is composed mostly of normal matter, just like the rest of the solar system, the positrons quickly disappear. Now we are left with a newly formed helium

nucleus and six gamma-ray photons, two of which originated from the
original reaction and four which are a side product of the positron transformation. Gamma-ray photons are a very energetic form of light, created by nuclear fusion. Thus, fusion transforms a tiny part of the mass
into a highly dangerous form of light. That light is now moving around
deep inside the solar core, but this is only the beginning of its journey: it
is going to be substantially transformed before it reaches the solar surface, from where it shines brightly as—mostly—ordinary visible light.

The density of matter in the solar core is very high. Even though it
is a gas, it is compressed so strongly that in the center of the Sun each
volume of gas is about 150 times heavier than the same volume of water
on Earth. The gamma-ray light created during nuclear fusion does not
get far through matter that dense before it collides with it. As it bounces
around between the gas particles, some of the energy in the gamma-
ray light is rapidly transformed into movement of those gas particles,
with light of less energy left over. The released fusion energy is thus
converted into a mix of particles that rapidly move about while colliding with their neighbors—which is saying nothing but that it becomes
hot—all bathed in the glow of light. The photons that make up that
light continually transform into the energy of moving gas particles and
back into photons. No individual photon survives for very long within
the solar interior, only the mix does, like drops of water in an ocean that
are part of the larger body of water and not discernable as individual
drops with any lasting identity.

Eventually, the fusion energy released in the solar core does leak
toward the surface, from where it escapes the Sun. Photons move, of
course, at light speed (300,000 kilometers or 186,000 miles per second), so
if they could fly unperturbed in a straight line it should take them just
over two seconds to travel from the solar core to the solar surface. However, because they have to travel in a random walk, scattering all the
time through the solar gas, the average speed at which nuclear energy
leaves the Sun is nowhere near the speed of light: every time a photon
interacts with a gas particle, the direction in which the energy travels is
deflected into another so that the net rate of progress outward is greatly
slowed. Because of that scattering, it takes energy something like 100,000
years to move out from the solar core to the surface, all the time transforming from photons into the energy which is the heat of the gas, and
back. The average propagation speed for energy from the core of the
Sun to the solar surface is thus about two tenths of one millimeter per

second, which is about a meter or yard per hour—and that is, surprisingly, even slower than a snail's pace.

In the innermost 71% of the Sun by radius, or 36% of its volume, light moves by continually scattering against, and sharing energy with, the gas. Further out, however, the gas cools down to a state in which it becomes less transparent, more like a fog than like a clear atmosphere, so that light has an even harder time moving outward. In that outermost shell of the Sun, energy is mostly transported by large-scale motions of the gas, much like the overturning motions we can see in a pan of soup on a hot stove. Here, energy is transported in the form of heat rather than in the form of light. In other words, it is transported as the motion of gas particles rather than as the motion of photons. A byproduct of this gas motion is that a very small fraction of the total energy is converted into electromagnetic energy of a different kind: the Sun's magnetism, causing sunspots and solar flares at the surface, is a direct consequence of the escaping fusion energy that is traveling through the opaque outer layers of the solar interior. We return to that, and to its consequences, in Chapter 12.

When the hot gases within the Sun rise to the surface, a final energy conversion occurs before that energy can reach the Earth: the gas particles transform their thermal energy into light, which, because it is now at the surface of the Sun, can escape into interplanetary space. The energy is now moving as light does everywhere: at a speed of close to 300,000 kilometers per second. After a long trip through the solar system it will reach interstellar space, unless, of course, something gets in the way that blocks the light's escape into the vast emptiness that surrounds the solar system. That something could be a planet.

Seen from the Sun, a planet is a very small object. This small size means that the fraction of sunlight intercepted by the Earth eight minutes after leaving the sun is tiny: only slightly more than one part in a million of the Sun's luminosity falls on the Earth. Of that, approximately 30% is immediately reflected back into space, most efficiently by clouds, snow, and ice, but also by the oceans and the continents. The remaining energy, on average 250 watts (or four typical incandescent light bulbs) per square meter, is absorbed in some form or other into the Earth's atmosphere, oceans, and continents, including any living organisms in or on them. The globally averaged temperature of the Earth is essentially constant, apart from a slight overall increase associated with climate change. To maintain this balance, the Earth must lose, on average, an amount of energy equal to what comes in from the Sun. Apart from the direct reflection of sunlight, most of the Earth's emission is in the form of infrared light, which is the invisible but noticeable glow of all warm objects.

The balance between incoming energy from the Sun and outgoing thermal glow maintains the globally averaged temperature remarkably well. A simple estimate of the temperature at which this balance settles can be made by comparing the solar energy input with what one would expect to be emitted at a range of Earth temperatures. As the temperature of an object, be it a planet or a rock or even a star, increases, it emits light that is brighter and that shifts in dominant color from infrared, through red, going by all the rainbow's colors, to eventually reach blue or—if hot enough—even further, into the ultraviolet and X-ray domains. Planets in the solar system are not warm enough to emit in the visible range (although they do reflect the Sun's visible light), but rather glow in the infrared, which we cannot see with our eyes but which our skin can feel as heat. The same fundamental physical law applies to infrared and to visible light, so we can use that general law to compute the temperature of a planet, given its distance from the Sun, as if it were only a simple glowing sphere, without an atmosphere or oceans. For such a desolate planet in an orbit like the Earth's, the average temperature would be about minus 18 degrees centigrade (0^0F). For our Earth, however, the actual average temperature is close to + 15 degrees centigrade (60^0F). For the planet Venus the difference between the actual and expected temperatures is much larger than that for Earth: Venus's surface temperature should be several tens of degrees centigrade, but instead it is a scorching 460 degrees centigrade (850^0F).

Why are the estimates so different from reality? Nothing is wrong with the physical equation used to make the above estimates. What is wrong is the assumption that it could be applied this simply: there is an atmosphere to contend with, and that changes things markedly. If the atmosphere is thin, the law works pretty well. For the planet Mars, for example, which has a very thin atmosphere, the average temperature is about minus 55 degrees centigrade (with large regional differences, just as on the Earth), which is close to what it should be in the absence of an atmosphere. What happens in the cases of Venus and the Earth is that there is an atmosphere around these planets that traps heat, so that both are considerably warmer on their surfaces than they would be without one. The blanketing function of an atmosphere is known as the greenhouse effect.

In recent decades, we have become very concerned about the greenhouse effect of the added carbon dioxide that our constantly growing, industrialized society puts into the atmosphere. If we had no greenhouse gases in the atmosphere, most of the Earth's oceans and landmasses would be covered by ice, and ice reflects incoming light more efficiently than the well-known blue–brown–white–green sphere that we live on, so that the global equilibrium temperature would drop to below minus 50 degrees centigrade. In other words, the entire Earth would be as cold as the south pole in the middle of its winter. If we had a much larger amount of greenhouse gases, on the other hand, climate could run away on us, as scientists think happened on Venus. The thick Venusian atmosphere consists mostly of carbon dioxide, which traps heat so efficiently that the temperatures near the surface by far exceed the average temperature expected in the absence of that atmosphere. Venus may have had water a long time ago, but the runaway greenhouse effect would have boiled any such surface water off the planet's surface into a dense, steamy vapor (thereby adding even more greenhouse gases to the atmosphere). High in the atmosphere of Venus, where it would be subject to the Sun's ultraviolet light, this potential water vapor would have decomposed into hydrogen and oxygen. Hydrogen would gradually escape, being stripped off the planet by the solar wind (which we will get to in Chapter 11), while the remaining oxygen would be bound to other chemicals, leaving Venus without atmospheric water, and with only an inhospitable, hot, dense carbon dioxide atmosphere after the first hundreds of millions of years of the planet's existence. This is indeed what we see today.

On the Earth, water vapor, carbon dioxide, and other greenhouse chemicals raise the temperature over the chilly value expected for a planet without an atmosphere. They increase the temperature just enough to remain within tens of degrees above freezing across most of the planet. This keeps the global ice cover fairly low, with the areas with most ice and snow restricted to the Arctic and Antarctic regions. At the present average temperature, the condition of the globe is such that an average of 70% of the energy contained in the incoming sunlight is absorbed into the Earth's atmosphere, oceans, and biosphere. That energy is eventually re-radiated into space as infrared light, but not before it powers our entire biosphere.

The energy contained in the sunlight shining on the Earth is converted into a wide spectrum of distinct types of energy. Much of the solar energy warms the top layers of the oceans, and from there it is slowly mixed into the deep layers. Air above the land and above the oceans is also warmed, mostly by heat radiating from the warmed masses of land and water. The combination of warm air, mixed with cloud-forming water vapor, on a spinning sphere that is subject to sunlight on only one side sets up the circulations that we know as weather patterns: part of the solar energy works to power the winds and rains across the globe. Winds and oceans couple through the exchange of heat and through the force that the winds exert on the oceans. This force interacts with the ocean surface through the waves that it causes. That coupling sets up the global patterns of the circulations in the oceans, leading to vast, slow eddies, and the sinking of surface waters into some of the deep oceans and their eventual rising again from there elsewhere on the globe. The Sun's energy shining on the Earth is thus carried through a chain of processes that occur even in the absence of living organisms. These processes redistribute that energy, which eventually is radiated into space as heat. The time scale on which this happens is as short as just a few hours for some process chains, being radiated as heat from the dayside of the Earth even as sunlight comes in, or a few hours later during the cooling that takes place throughout the night. For other process chains it can take amazingly long, over 1000 years, as in the case of energy participating in the slow circulation of water through the deepest oceans.

Some animals use sunlight, converted into heat, to enable their movement. Reptiles, in particular, have to rely on sunlight to warm them up sufficiently so that their bodies' chemical systems function efficiently enough to allow them to extract chemical energy that powers their muscles so that they can hunt or evade other hunters. Most fish rely on warmth from the ocean waters to maintain their bodies at an appropriate temperature for the same reason, with some exceptions such as the warm-blooded fast-hunting tuna. Most animals, including humans, do enjoy warming up in sunlight well enough to frequently bask in the Sun for a little while. In the end, however, all animals rely on plants to capture sunlight for them (except for tiny fractions that are used directly, for example to create vitamin D). It is really the digestion of sunlight captured by plants that enables animals, including us, to live.

Plants and many types of bacteria use the energy in sunlight to grow. The first step by which they do this is by using sunlight to power chemical reactions that convert water from the ground and carbon dioxide from the atmosphere into sugar. The oxygen that is released as a byproduct of these chemical reactions is expelled from the plants into the atmosphere. In effect, plants do the opposite of what animals do: animals breathe in oxygen and exhale carbon dioxide as they burn sugars, while plants breathe in carbon dioxide and breathe out oxygen as a series of chemical reactions enables them to store the energy of sunlight within the sugar molecules that they create. The set of reactions in plants that achieves the chemical capture of solar energy is called photosynthesis. One key step in that process involves chemicals that are called chlorophylls and carotenoids. Chlorophylls and other pigments absorb the red and the blue parts of sunlight, reflecting the rest back from the plant. The reflected part of the sunlight thus lacks red and blue relative to the incoming sunlight, which is why we perceive chlorophylls and most plants and trees containing it as green (not the pure greens that are part of the rainbow, but a mixture of colors that our eyes and brains convert into various shades of generally darker, richer greens). Carotene, another pigment acting in the photosynthesis of some organisms, absorbs primarily blue and green, leaving more red light to be reflected off the organisms than off organisms with chlorophylls, resulting in colors dominated by an orange tint.

How plants managed to grow had puzzled farmers and scientists alike for millennia. If was, of course, a long-established fact that plants need water to thrive. It was assumed that plants grew by using their roots to absorb water and similarly to assimilate soil as food to enable growth. In the mid-seventeenth century, Jan van Helmont, a Flemish thinker, performed an experiment that showed that plants, in fact, use very little of the soil in which they are rooted. For five years, he cared for a willow tree growing indoors in a pot, all the time carefully measuring how much water he gave it. At the end of that period, he noted that the weight of the soil in which the now much bigger tree was growing was essentially the same. He concluded that the plant had grown by somehow converting water into wood, bark, and leaves, not yet knowing that atmospheric gas was also involved.

It was not until 1727 that Stephan Hales wondered if light might have something to do with the growth of plants. Concepts of the involvement of components of air developed slowly, and it took until the end of the nineteenth century to establish our knowledge of the different chemicals in the air. Joseph Priestley, a British researcher, demonstrated in 1772 that whereas a candle would stop burning after a while when placed inside a glass jar, adding a plant to the jar could keep the candle going much longer. The same was true, he demonstrated by a rather cruel experiment, for plants and mice. He also noted that daylight caused bubbles to be exuded by green organisms living on the walls of a glass jar of pump water. When Priestley informed Antoine Lavoisier, a French chemist, of his work, Lavoisier realized that Priestley's gas was one component of air; he named it oxygen.

Jan Ingenhousz, a Dutch physician, connected the above observations when, in 1779, he published work showing that the positive effects of a plant on the quality of air for breathing by animals and people are limited to periods of daylight. He carefully distinguished in his experiments between the roles of warmth and of light, noting that it was only light that caused plants to transpire the gas that kept flames and mice alive. Ingenhousz saw a big picture emerge and wondered whether plants might not be in effect cleansing the air, thus making it usable by animals—as indeed they do.

By the end of the eighteenth century, the basics of photosynthesis were known: sunlight enables water and carbon dioxide to be taken up by plants, which then release oxygen gas as a byproduct of sugar production. The details of the photochemical reactions and the role of various parts of plants in the reaction chain, however, continue to be studied for the diversity of plants and bacteria that are capable of photosynthesis in some form or other.

The sugars and other molecules that are made in the process of photosynthesis may be stored somewhere within the plant or tree for later use, or they may be used immediately for the creation of new tissues, such as leaves, stems, bark, or roots. Whenever sugar or other energy-storage chemicals are used for growth, oxygen is re-captured from the atmosphere in order to burn chemicals in the plant to release energy and to thereby enable the assembly of materials needed to build the plant. Storage of solar energy occurs in the form of carbohydrates (including sugars and starches), oils, and proteins, all derived from processes starting with photosynthesis. The waste product of photosynthesis, oxygen, enters the atmosphere. Approximately half of the oxygen currently entering the atmosphere is being created by trees, shrubs, plants, and grasses on land. The rest of it is produced by single-celled plants in the oceans, called phytoplankton.

Humans obtain their energy by consuming energy-carrying chemicals. They do so either by eating plants, or by eating animals that eat plants (herbivores), or by eating animals that eat animals that eat plants (carnivores), and so on. In the end, all the energy used inside the human body derives from plants that obtained it from sunlight themselves. This energy is extracted by animals by combining these energy-carrying substances with oxygen, thus in effect burning the plant materials to enable the activities of life, such as growth, movement, and thought.

All of the energy that we obtain from consuming or otherwise using plants, animals, and fossil fuels was captured as sunlight some 100,000 years after nuclear collisions in the center of the Sun turned mass into light. The energy yield from nuclear reactions in the Sun is astounding: to power a human over an entire life span (of 70 years as an example, using the current world average of 2800 calories per day) requires the fusion of only half a gram of hydrogen into helium. The total chemical energy stored in an average human being, mostly stored in the chemical bonds of our sugars, fats and proteins, is about 110,000 calories, equivalent to the yield from 0.75 milligrams of fused hydrogen. Einstein's equation, $E = mc^2$, shows that the energy of that stored sunlight adds merely 5 micrograms to the weight of the atoms in our sugars, fats, and proteins.

We now turn from the energy captured by photosynthesis to the chemical aspects of that process. Photosynthesis in plants consumes carbon dioxide and releases newly created oxygen gas into the atmosphere. If plants were to die and decay on the planet's surface, the oxygen that any particular plant produced is theoretically consumed again by the plant's decay, releasing the carbon dioxide that the plant had consumed during its growth. If chemical decomposition on the surface were complete, no net change in carbon dioxide or oxygen in the atmosphere would be seen when comparing times before and after the plant's existence. The decay of plants is rarely complete though. Wherever plant materials are taken out of reach of the atmosphere before they fully decay, some or all of the released oxygen remains in the atmosphere and little carbon dioxide is released by the incomplete decay. This happens commonly when land plants or their remains end up covered (thus forming thick layers of soil), when phytoplankton sinks to the bottom of the oceans and seas, or when dead animals end up similarly out of reach of the atmosphere.

Geological processes can cause vast masses of plant and animal materials to be buried for very long times. In the process of being increasingly buried under layer upon layer of sediment, the organic material can change into coal, oil, or gas. Plant and animal deposits from the oceans and the seas predominantly form oil, whereas land plants tend to predominantly form coal. In the Earth's history, abundant life in the oceans occurred well ahead of substantial life coverage on dry land. It is therefore no surprise that oil can be substantially older than coal. The average age of oil fields is approximately 35 million years, but some (such as those in eastern Siberia) are approximately 600 million years old. Some coal fields are close to 400 million years old, but it was around 250 to 320 million years ago that grasses and other plants (at first dominated by ferns, later including shrubs and trees) were abundant enough to begin to create substantial coal fields. It was in this era (the Permian and Carboniferous Paleozoic eras) that most of the black coal on Earth was formed. A second main black coal-forming period occurred from 65 to 200 million years ago in some parts of the world. More recent eras have created most of the brown coal (lignite) but they also created black coal in different geological settings. So, depending on where the coal is being mined, it may be anywhere from 10 million to some 320 million years old.

Mankind has used oil and coal for millennia. Herodotus reported on the use of an early form of asphalt in construction projects in ancient Babylon, four millennia ago. Oil wells were drilled in fourth-century China. With the growth of the Earth's population and the range of applications, oil and coal use increased rapidly from the Middle Ages onward. By the fifteenth century, bricks had been invented, enabling, among other things, the relatively cheap construction of chimneys, which, together with the growing importance of cities, resulted in a growing coal market. Natural gas saw its first major use in lighting, as did petroleum when it started to displace whale oil as lighting fluid. By the 1860s commercial oil wells were being drilled in the Western world. Then, in the early twentieth century, when automobiles and electricity became available to growing numbers of people, oil, coal, and gas usage grew ever more rapidly. Current use of all fossil fuels is close to 4 billion tons of oil per year, plus some 2.5 billion tons of oil equivalent in the form of coal, and almost that much again in the form of natural gas.

That amounts to the equivalent of more than a ton of oil per person per year. By the way, incoming sunlight provides 12,000 times as much energy, which is a quite convincing number to show that solar energy is easily able to supply mankind's energy needs, if it could be efficiently harvested.

The increased use of fossil fuels, combined with deforestation and other human activities, causes the carbon dioxide content of the atmosphere to gradually increase. At the time of this writing, there is about 40% more carbon dioxide in the atmosphere than in the pre-industrial era. It is a greenhouse gas, which means it causes heat to be retained by the atmosphere. We understand to some extent that the consequence of this increase in carbon dioxide levels is that the average temperature of the entire globe rises. How far this will go in changing climate, sea levels, and—eventually—the biosphere is an active field of study. The atmospheric–oceanic–geological feedback system resulted in a fairly stable climate for millions of years, albeit irregularly punctuated by ice ages. These days, however, the biosphere has a new component: mankind's economy. Changing climates will change the water cycle and the sea levels, the crop yields, and the forestation levels. The deep oceans form a vast, slowly responding reservoir for heat that will follow the atmospheric and shallow-waters temperature changes on time scales of a century or more. The human population and mankind's economic activities are changing even as the balance between all these coupled reservoirs and activities is adjusting. The consensus among climate scientists is that global change is happening and that it will persist for at least some decades to come. However, a reliable prediction of what happens after that, including where a new balance, if any, will be found once we actually acknowledge what science tells us and change our ways, is yet to be achieved.

These various astronomical, geological, and biological processes provide us with but a glimpse of the astonishing diversity of the source materials that make up our bodies. The story of carbon alone is amazing. Some two-thirds of the carbon in our bodies is coming to us from the ongoing cycle of photosynthesis of growing plants on land and phytoplankton in the seas, or the consumption of these plant products by animals. About one-third has been locked up in subterranean oil, coal, and gas deposits for millions to hundreds of millions of years, having been released from there sometime over the past century, with most of that release occurring in the past few decades. That release occurs through the burning of fossil fuels, which results in newly created atmospheric carbon dioxide, which carries the carbon into the atmosphere from where it cycles into plants, animals, and thereby—eventually—into our bodies as we consume our foods. The use of fossil fuels thus mixes long-buried carbon into our biochemistry. A fraction of the carbon atoms in our bodies has been locked up in oil since the development of the first aquatic animals, some 500 to 600 million years ago. Other carbon atoms were buried in the Earth since the early development of land plants some 475 million years ago, or after having been part of the earliest insects, amphibians, and reptiles over the next 100 million years. Carbon then participated in the emergence of mammals, birds, and flowers over the next 100 million years, and was part of the large dinosaurs, which became extinct 65 million years ago. Mankind has not really been on this planet for very long, so that carbon released by us into the atmosphere from the combustion of oil and gas is unlikely to ever have been in a human before it was excavated or drilled out of its storage areas sometime during the past century, years, or weeks.

Our bodies rely on components of chemicals as well as on the energy stored in chemicals. Derived from plants and—unless we commit to purely vegetarian diets—from animals, the energy and building blocks are used to power the body and to maintain it. We have seen a glimpse of what elemental ingredients these contain, and we touched on how these ingredients cycle through the biosphere and the Earth's interior. There is much more to these stories, however. First we look at the chemical elements that make up our bodies. Then we look at how these elements move around us on time scales from days to hundreds of millions of years.

Key Points: Chapter 5

- The ultimate source of energy for virtually all life on Earth is sunlight, which originates from nuclear fusion reactions within the Sun.
- It takes approximately 100,000 years after nuclear collisions in the Sun's core that turn mass into light for that light to leave the solar surface. The vegetables and fruits that we consume today have incorporated the energy from light that left the Sun a mere eight minutes earlier. From then it may take only weeks before that plant (and with it the Sun's energy) enters our bodies.

- On average, 70% of the incoming sunlight is absorbed into the Earth's atmosphere and collective ecosystems. Eventually, that energy re-radiates into space as infrared light, but not before it powers our entire biosphere.

- Some two-thirds of the carbon in our bodies is derived from the on-going cycle of photosynthesis of plants and our direct or indirect consumption of these plants. The rest comes from atmospheric carbon dioxide created by the burning of fossil fuels and other processes, from where it cycles into plants, and eventually into our bodies.

- Almost all of the natural energy used on Earth, from food to fossil fuels to the renewable resources that rely on moving air or water, is in fact transformed sunlight. Nuclear energy and geothermal power are the only exceptions.

- **Astronomical, geological, and biological processes account for an astounding diversity of molecules in our bodies.**

The world of chemical reactions is similar to a stage, on which scene after scene is played out without pause. The actors on it are the elements.

CLEMENS ALEXANDER WINKLER (1838–1909),
in *Naturwissenschaftliche Rundschau* (1897)

6

The Human Elements

Everything is subject to wear and tear. To maintain the functionality of anything over time, things that wear out, break, or are lost need to be replaced. This is as true for a car as it is for a city or for an entire continent. The need to repair and replace components is also true for a living organism. All of these structures need materials to replace parts that wear out, to power parts that move, and—for warm-blooded organisms such as we are—to supply energy to keep them warm.

When we take care of a car, for example, this is done most effectively if we have ready access to replacement parts, have nearby mechanics with experience working on the make that we own, and have the right kind of fuels available within a reasonable distance. Similarly, for a vibrant, growing city, building materials, food, water, and energy need to be readily obtainable and transportable into the city's infrastructure. Some things can be brought in from afar nowadays, of course, so scarce car parts or desirable food items can be obtained, but there is a price to pay to obtain things that are regionally unavailable while a there is a risk that sometimes the desired articles cannot be obtained at all.

Even entire continents age and need rebuilding to survive for hundreds of millions or even billions of years: their erosion is counteracted by the geological activity that forms mountain ranges where continental plates collide or by the volcanic activity that spews new materials onto the eroding landmass. This may happen locally on the eroding structure itself, or the locations where new material is deposited may shift around with time from one location to another on a continent. However it happens, the net effect is that landmasses on the Earth are maintained despite billions of years of steady decay.

Living organisms need building materials and energy for their structure and function. Timely availability is particularly critical when they are in their growth phase, but even mature organisms continually require chemicals to power them as they burn through reserves and need mixtures of other chemicals to replace cells that die. Modern-day human beings are fairly effective in acquiring sustenance even over substantial distances, but they benefit from being able to use what is readily available locally. Animals can hunt within their domains and thus have a reasonably large reach over which they can collect what is essential to them. They have the ability to seek out particular nutrients that might exist in only certain locations, such as nectar in flowers or salt in concentrated deposits. Plants, on the other hand, have to rely on what is right there, where they first established themselves, having access in addition only to what may be carried to their location by wind, water, and animals. Early, simple life forms would have had to rely on what would be brought into contact with them by water or wind wherever they were. If there was an abundance of what they needed, they could thrive. If their needs were not met, however, they would die. A resilient primeval species, one that could most readily withstand natural selection associated with the limited availability of resources, would thus evolve to be rather tolerant of temporary surpluses or shortages of resources while making optimal use of the most commonly available ones.

Looking at it from an evolutionary perspective, it is no surprise that living organisms are primarily composed of what is most common at the Earth's surface. With the oceans covering two-thirds of the globe's surface, it is quite natural that water makes up the bulk of any living being. The chemicals that are readily available to a given organism depend, of course, on its surroundings. Over time, living beings have adapted to regional differences. Plants, for example, are differentiated by their affinity for alkaline or acidic soils, and differ in their tolerance of either abundant water or lasting droughts. Most animals are tuned to particular diets, which can range from plants or their nectars to a variety of other animals.

Despite these variations, the bottom line is that, by and large, living beings are composed of what is within easy reach around the surface layers of our planet. This is immediately obvious when looking at the mix of elements out of which the Earth is made. An element is a chemical that is so simple that it cannot be reduced to a blend of simpler chemicals, neither by interactions with other chemicals nor by subjecting it to heat. There are 98 elements that occur naturally on Earth, and another 20 or so that can be artificially made but that do not survive long enough to exist in a natural setting on the Earth in any measurable quantity. Of the 98 naturally occurring elements, only 21 seem (at present) to be essential to human life.

In contrast to the elements that directly support human life, the variety of chemicals involved in life on Earth is staggering: the variety of molecules that comprise a living organism contain anywhere from just two atoms to over 10 billion atoms, and one can make an incredibly large number of molecules by varying the number, mix, and order of atoms within them. Living organisms are generally not exposed to elements in a pure form, but rather encounter them as chemical mixtures (such as oxygen in air) or bound with other elements into more complex chemicals (with hydrogen and oxygen forming water molecules among the most common ones). Yet, rather than reviewing an exceedingly long list of chemicals that we need to function, we can learn much by starting with the very much shorter list of elements that are essential to life.

Modern physics and chemistry laboratories have equipment with which it is rather straightforward to find out whether a chemical is a pure element. Most of the elements were identified as such about a century ago, however, when equipment and understanding were far less advanced. For a long time before that, experimenters and philosophers believed that all matter, in all its rich diversity, was created from four or five ingredients, which were referred to as elements but which we now do not recognize as such. The earliest formalized thinking about elements considered water, fire, earth, and air as the basic building blocks of any substance on Earth. Later a fifth element, referred to as ether or quintessence, was added to describe what supposedly made the things in the heavens.

Interestingly, people had found materials that were nearly pure elements, or had refined chemicals into nearly pure ones, long before they recognized them for what they were. Among these were copper, lead, gold, iron, and carbon (both in the form of charcoal and as diamonds), all of which have been worked with for well over 5000 years. The modern understanding of chemistry and its elements is far less old. It was not until 1789 that Antoine Lavoisier published a book in which he first presented a systematic summary of what were then thought to be the "simple substances", those that could not be further divided, or at least not with the means available at the time. Without the high-tech means available to us now, it was a difficult undertaking to understand substances and to differentiate between the elemental and composite ones. Lavoisier was among the first to exhaustively compile the available evidence for the existence of a substantially longer list of "elements" than the four or five considered since classical times. Even he recognized, however, the difficulty of knowing whether any substance was really in its simplest form. The atoms that make up the elements are indeed the smallest units into which matter could be divided in Lavoisier's time; further division was not possible until, well over a century later, nuclear physics came into play.

Lavoisier's list included 33 substances, most of which we now indeed recognize to be elements. His list also included light and something that he called "caloric" (by which he meant caloric heat, or, as Lavoisier described it more precisely, the cause of heat). These two attributes are not elements at all by the present-day definition, but were reasonably included given Lavoisier's definition. Even he realized, however, that caloric was an odd entry to include: he could not, for example, be sure whether it was different from light. Given the technologies that he had available, he could not contain caloric in a vessel and he could not determine if it had any weight. It even seemed to work as a repulsive medium. Indeed, heat causes substances to expand and, with enough of it, it can even break up chemicals. Nowadays, we understand heat to be energy rather than matter. Light, as we saw before, is but one form of energy, so from the modern perspective it is no wonder that Lavoisier could not understand the similarities or differences between light and what he called caloric. Lavoisier's confusion as to the properties of energy relative to those of the elemental atoms was reflecting but one of the many difficulties in identifying among all the known chemicals those few that were the building blocks of others. In the case of energy this was especially difficult, because it is clearly involved in the making and breaking of compound chemicals. Our terrestrial intuition sometimes suggests distinctions that turn out not to be useful or correct in terms of physics. Nowadays, with light and heat much better understood, the definition of an element excludes energy because it is not a composite of electrons, protons, and neutrons.

Pure chemical elements are composed of atoms that are identical in one respect: each of those atoms has the same number of electrons orbiting nuclei with a matching number of protons. Electrons are negatively charged, while protons—almost 2000 times heavier than electrons—have the same magnitude of charge but of the opposite sign. Each atom is therefore generally electrically neutral, with its negative electron charges and positive proton charges adding up to zero. In many circumstances that we shall encounter later on, atoms may lose one or more electrons, so that they are no longer electrically neutral, but, in isolation and at temperatures that we are most comfortable with, they do tend to be.

The nuclei of the atoms comprise a mix of protons, each with a unit of positive electrical charge, and neutrons, which are electrically neutral—which is what led to their name. Neutrons act as a kind of glue that keeps the protons from flying apart subject to their own mutually repulsive electrical force. The number of neutrons roughly matches the number of protons in a core. Yet, even atoms of the same element can have slightly different numbers of neutrons. Hence, physicists use the number of electrons or the matching number of protons of an atom to sort the elements into individual species. That is perfectly suited also for chemists and biochemists because it is the number of electrons in the atoms rather than their atomic mass (determined by the numbers of both protons and neutrons) that determines an element's chemical properties.

The properties of the electron clouds orbiting the atomic nuclei became known only when the theory of quantum mechanics was developed in the first half of the twentieth century. Quantum mechanics enables us to understand the physical reasons why elements can be grouped according to their chemical properties. Yet, that grouping by chemical properties was known already well before the development of quantum mechanics. In 1869, Dmitri Mendeleev published what we now know as the periodic table of the elements. At the time of his work, 63 elements had been identified, and Lavoisier's light and caloric had already been taken off the list. Mendeleev, following less successful earlier attempts of his colleagues, managed to sort the known elements according to their chemical properties and atomic weights. His tabulation was so successful that gaps in it turned out to predict the properties of the 35 elements that had yet to be discovered. Even though Mendeleev's original table was incomplete, it already included all of the elements needed in the human body. In fact, those 21 elements were already known by 1817, when the last member of that the list, selenium, was discovered. It took a long time before at least some of the functions of the elements in the human body were known, however, and many of their functions remain to be identified even today.

Sorting the elements that have known functions in the human body by the number of electrons in their atoms—known as their atomic number—reveals that the highest atomic number in that list is number 53, iodine. But number 53 is more than twice the number of elements that are known to play a role in the human body. Apparently not only are the heaviest elements beyond number 53 unused by human chemistry, but more than half of the ones with atomic numbers below that of iodine are also excluded. Can we understand why?

To offer at least a partial answer to the question why only about one in four of the chemical elements that occur naturally on Earth are involved in human biochemistry, we can look at the availability of the elements for use in chemical reactions. We can divide that perspective into two parts. One part addresses availability in general. Before we talk about that, however, we first look at the more restrictive part of that perspective which involves the ability of elements to participate in chemical reactions. Chemical reactions are interactions between atoms in which they share electrons. For reasons based on quantum mechanics, atoms prefer to be in states with particular numbers of electrons around them. Interestingly, the optimal number of electrons for atoms is generally not the same as what they naturally need to be electrically neutral, which is set by the number of protons in their nuclei. The consequence of this pair of different requirements is that atoms prefer to partner with atoms of particular other elements and sometimes with atoms of their own kind, so that the total number of shared electrons is optimized. As a result of such partnering they can share their electrons in such a way that the electrons around each of them are in the quantum-mechanically perfect arrangement, while their groupings remain electrically neutral.

Among the 98 naturally occurring elements, some actually have the number of electrons that balances their electrical charge and that meets their quantum-mechanical needs at the same time. The atoms of such elements do not need to share their electrons with anything else. In fact, it is extremely difficult to make them do so. These elements react chemically only under very special circumstances, and then with great difficulty, even in reactions with their own kind: they are effectively chemically inert. And because they are inert, they can play no role in the chemistry of life. These elements are what we refer to as the noble gases, called "noble" precisely because they are influenced by virtually no other chemical around. Thus we can understand why elements number 2, 10, 18, 36, 54, and 86 do not play a role in our bodies. These are the noble gases helium, neon, argon, krypton, xenon, and radon. Their inability to partake in any chemical reactions made them difficult to identify. Consequently, all but one of them were discovered relatively late, between 1894 and 1898. The noble gas that was first discovered to exist, in 1868, was helium. Helium is the second most abundant element in the universe, second only to hydrogen. Yet it is very rare on Earth. It is very light (only hydrogen has a lower weight) and does not bind to anything else that could make it part of a heavier molecule. Consequently, almost all the helium has vanished from the Earth, evaporated from the top of the atmosphere over the billions of years of the planet's past.

There are only four noble gases with atomic numbers below that of iodine, which has atomic number 53. This leaves us with 28 elements with atomic numbers below iodine that are not known to play any role in the chemistry of our lives. For most of those, it may well have been their general availability in nature that has made them less than suitable to be incorporated in biochemistry.

Determining the general availability of the elements to living organisms involves a hierarchy of processes. First, we have to know the chemical makeup of the cloud of interstellar gas out of which the solar system was made. Then we have to understand how that cloud made planet Earth. And finally we need to look at what the physics and chemistry of the Earth—including that of life itself—have done to the availability of chemicals over 4.6 billion years of its history. Later chapters go through many aspects of those issues. A bird's-eye perspective, even one that necessarily leaves out many of the complex details, gets us a long way to understanding the elemental chemistry of life, however.

The composition of the solar system, out of which the Sun, planets, comets, and interplanetary dust were formed, is now known quite well. Obtaining this knowledge has taken many years and rests on the utilization of a multitude of methods. Given that we are largely bound to the Earth, the first bits of that knowledge came from studying the Earth itself and any object that fell onto it, namely the meteorites. Most of the mass of the solar system is contained within the Sun, however, and traveling to the Sun remains beyond our technical capabilities: the Sun is so hot that anything we can build would melt and evaporate long before it could reach the surface for an analysis on the spot. Thus, although spacecraft can now at least sample the wind blowing off the Sun and scoop up some of the interplanetary dust, astronomers mostly use a surrogate method to learn about the chemical makeup of the Sun and of other bodies in the universe. The study of light by a method called spectroscopy is as fundamental to astronomers as the use of a stethoscope is to a physician.

The electrons orbiting the atomic nucleus that determine the chemical properties of elements also determine the physical properties of the interaction of matter with light. Light propagating through a cloud of gas, such as through the Sun's atmosphere, encounters the electrons of atoms that are part of that cloud. These electrons orbiting atomic nuclei within the cloud can absorb or emit particular colors of light. When light is unraveled into its constituent colors, such as by a simple prism or by a spectrograph, its high-tech equivalent, each chemical compound is seen to take or add light in a series of particular colors. Thus, a detailed study of the colors of the light from any astronomical body, including the Sun, tells us what chemicals exist in the material through which that light has traveled. Further, just how much light of each color is removed or added tells us how much material of each type there is. This is by no means a simple experiment, but it does allow us to establish what the Sun is made of without going there. In fact, it enables astronomers to measure the composition of any star throughout the observable universe.

A comparison of the colorful fingerprint of elements in light, as it can be studied in a laboratory, with the patterns seen in sunlight enabled the determination of the composition of the Sun, in which the bulk of the mass of the solar system resides. One of the early surprises in solar spectroscopy was the discovery of an atomic fingerprint in the Sun's light that had no counterpart in known laboratory measurements at the time. In 1868, observations of the Sun revealed the existence of a chemical that was unknown on Earth. Because of its prevalence in the Sun, two British scientists, Norman Lockyer and Edward Frankland, proposed naming that new chemical after the Latin word for the Sun: helios. It took more than two decades before the existence of this element was confirmed on Earth. We now know it as helium. Helium accounts for 23% of the total mass of the universe if we look only at "ordinary" matter and exclude the so-called dark matter (and in fact we have no idea what that might be). Helium remains the only element first discovered outside the Earth by astronomical spectroscopy, but spectroscopy in laboratories has been a very successful tool in the discovery of many other elements of which only relatively small amounts exist on our planet.

When comparing the list of abundances of elements in the solar system with those in the human body, we can draw a fairly straightforward conclusion. Rather than look at the chemical properties or the atomic number, we should be looking at the availability of the element: the human body uses essentially everything that is more abundant in the local universe than copper (with atomic number 29), provided that it can be involved in chemical reactions. There are only five exceptions to that rule: aluminum, silicon, and titanium—each more common than copper but not obviously essential to human life—and molybdenum and iodine, which are less common than copper but are part of human biochemistry. There is one other exception if we look not only at human biochemistry but also that of plants: selenium is used by mammals but is not part of the biochemistry of plants. In summary, looking only at availability for chemical reactions, human life—and, in fact, most other life forms on Earth—ended up using all but three of the elements at least as common as copper, incorporating only molybdenum and iodine as exceptions less abundant than copper.

Whereas the above argument makes for a relatively easy summary of elements used in human biochemistry, there is more to the story. The mixture of the elements at the Earth's surface is not simply a reflection of what occurs in the local universe. Many of the lighter elements and lighter chemicals have largely evaporated off the Earth in the past billions of years, much like helium has. Others have been preferentially incorporated in the mixture of metals in the Earth's core or in the rocky materials of the Earth's crust. Some make chemical compounds that readily dissolve in water, so that early life, which likely formed in watery environments, could readily access and process those elements. The many processes that are involved in determining the availability of the elements in different geological settings around the globe form a very long story that does not change the bottom line: life mostly uses what is readily available, while making use of just a few other elements that function in some particularly useful way.

Leaving out the inert noble gases, the most common elements in the solar system, in decreasing order, are hydrogen, oxygen, carbon, and nitrogen. Together, these four elements add up to 99.6% of all atoms in the human body (and nitrogen actually plays but a small part in these terms). In fact, these four elements are the foundation of all of the complex chemical compounds involved in organic chemistry. After these major four, there are seven that occur in relatively large numbers. These are, again in decreasing order, calcium, phosphorus, sulfur, sodium, potassium, chlorine, and magnesium. These 11 most common elements in the human body add up to more than 99.99% of all of our atoms. Interestingly, 10 of these are among the 11 most commonly found in seawater. This may reflect the fact that life likely originated in a salt-water setting. Only bromine, fairly common in seawater, is not used in human biochemistry, or by any mammal, for that matter. It is, however, part of the metabolism of non-mammalian marine life, so somewhere during evolution, bromine became incorporated in one but not in another branch of life.

The abundant elements in the body fulfill a host of different functions. Calcium is most common in our bones and teeth, but also plays a role in muscle action and enzymes. Phosphorus and magnesium, while also part of our bones, are both key to the pathways of use and storage of energy in the body. Sulfur helps shape both amino acids and the proteins that incorporate these acids. Sodium, potassium, and chlorine—in the form of dissolved salts—regulate fluid balances and the electrical couplings essential in the functions of the brain and the pervasive network of neurons throughout the body.

After the most common 11 elements follow 10 trace elements: iron, zinc, copper, nickel, manganese, iodine, selenium, molybdenum, cobalt, and chromium. Their known functions include oxygen transport (a crucial function involving iron), and enzyme and hormone functions. Various life forms use these elements differently. For example, whereas iron is important for oxygen transport through the bodies of most animals, some invertebrates (mollusks, such as the giant octopus, and arthropods, such as the tarantula) use a protein in which copper takes on the key role of binding to oxygen.

Life is not very tolerant to elements that it does not readily incorporate into its biochemistry: many of the elements that are not in the list of 21 with known human biochemical functions are toxic when concentrations substantially exceed the average over the Earth's surface environments.

Why life's chemistry does not appear to involve more of the relatively common elements remains subject to study. For example, aluminum is a very common element in the Earth's crust, yet appears to be unused in biochemistry. Because it is so abundant, it is perhaps not surprising that it exists within the human body, without toxic effects, in a fairly high concentration, even if somehow it did not find its way into the chemistry of life.

There is a greater variety of elements in the body than the 21 or so for which biochemical roles have been identified. This is hardly surprising, given that we live in a world in which all of the elements exist in a mixture. Some may fulfill one function only; for instance fluorine is known to prevent tooth decay but does not appear to have an essential role in human metabolism. Others may have a function even though it remains unclear what that is: deprivation of lithium, boron, silicon, vanadium, and arsenic (although a powerful poison in even tiny doses) has been attributed as the cause of some medical conditions. Some elements may be in the body with no real function at all.

Discovering the functions of elements that occur in very low concentrations within the body is difficult not simply because these elements are very rare, but because none of these elements is ingested in isolation. Salts, dissolved in water, offer the simplest molecules that enter the body. Some come indeed essentially as the atoms of their species: potassium, sodium, and chlorine, for example, mostly enter the body dissolved in water with only an electron more or less than they should have in their pure elemental form. Most elements, however, enter in the form of molecules in which atoms of multiple elements are grouped. Molecules can be as small as a pair of atoms, while others can contain hundreds or even many millions of atoms. Each type of these molecules is involved in the vast network of chemical reactions that starts after ingestion and continues until the waste products leave the body.

Very few of the compound chemicals that we need to sustain our bodies would exist in the absence of life on Earth: our lives are intricately coupled to the life of all plants, animals, fungi, and bacteria on the globe. The most common substance that we take in, and that makes up

most of our bodies, is water. Water is one of the few chemicals essential
to human well-being that would exist in the absence of life. Indeed, it
appears that life could not have developed without it. However, almost
all substances other than dissolved minerals in the water that we drink
or take in through our food exist because life forms other than our bodies created them.

The simplest of these molecules that is most critical to animal life is
supplied by plants: oxygen. This humble molecule, containing a pair
of oxygen atoms sharing their electrons, does not last long in the universe: oxygen reacts with a multitude of chemicals resulting in their
oxidation, which causes the disappearance of free oxygen molecules. It
is the metabolism of plants that creates oxygen as their waste product
that, in turn, enables animals to exist. Animals, including humans, rely
not only on the plants' oxygen, but also take nutrients from plants in a
vast diversity of molecules. All of these elements cycle through the biosphere and through the Earth, caught in biochemical and geophysical
cycles that we move on to next.

Key Points: Chapter 6

- A total of 98 elements occur naturally on the Earth. Of these, about 21
  are essential in supporting human life.

- Molecules in living organisms may contain as many as several million
  atoms. These molecules are astonishingly diverse, and this diversity
  is created by varying the number, mix, and order of atoms of which
  they are composed.

- Leaving out the inert noble gases, the most common elements in the
  solar system are hydrogen, oxygen, carbon, and nitrogen, which also
  form the foundation of the complex chemical compounds involved
  in organic chemistry. The 11 most common elements add up to more
  than 99.99% of all the atoms in our bodies, and closely reflect the composition of seawater.

- Very few of the chemicals that we need to sustain our bodies
  would exist in the absence of life on Earth. Water and carbon dioxide are exceptions, and life as we know it could not have developed without them. Almost all the other chemicals, though,
  apart from the small amounts of dissolved minerals that we take

in through food or water, exist because life forms other than our bodies created them.

- The elements that are used in the chemistry of our bodies and life on Earth closely reflect what is readily available in a chemically useful form, and are the ones most abundant in the solar system.

We are thus led to see a circulation in the matter of this globe, and a system of beautiful economy in the works of nature. This earth, like the body of an animal, is wasted at the same time that it is repaired. It has a state of growth and augmentation; it has another state, which is that of diminution and decay. This world is thus destroyed in one part, but it is renewed in another.

JAMES HUTTON (1726–97),
in *The theory of the Earth, Vol. II* (1795)

One cannot step into the same river twice.

HERACLITUS OF EPHESUS (535 BCE–475 BCE)

7

Cycles of Change

"Terraforming" is a relatively new word. It was first used in the latter half of the 1970s, and is still making its way into dictionaries. It is a well-known word, however, in the realms of science fiction and astrobiology, both of which focus on the potential existence of life on planets other than the Earth. Terraforming is defined as "planetary engineering designed to enhance the capacity of an extraterrestrial planetary environment to sustain life", as an action to "alter the environment (of a celestial body) to make it capable of supporting terrestrial life forms", or "to transform (a landscape on) another planet into having the characteristics of landscapes on Earth". The emphasis in these definitions lies explicitly on the activity being applied to some extraterrestrial environment, while the implicit assumption is that it refers to life as we know it on the Earth.

A long time ago, however, the Earth was an entirely different world that we would not recognize as being the planet we live on now. The Earth itself had to be terraformed in ways that continue to this day. During the first half of the Earth's 4.6 billion year history, there was very little free oxygen in its atmosphere, there were no plants or trees on the continents, and no animals in the seas, on land, or in the air. Single-celled cyanobacteria, however, did populate the early Earth: they came to be when the Earth was between 700 million and 1 billion years old. It is hard to establish exactly when life first appeared because the evidence is subtle and becomes increasingly indirect as we look further back in time. For relatively recent eras, we find fossilized animal skeletons, petrified trees, or even animal footprints and worm tracks. Early life, however, left very much less by way of evidence to go by. Microfossils that give us a view of the earliest single-celled life forms date back a little over 2.5 billion years. Fossilized structures that suggest that life existed span another billion years before that, but when looking more than 3.9 billion years ago, the evidence is limited to compounds that are associated with living organisms rather than to remnants or imprints of those organisms themselves. This indirect evidence suggests that life may have emerged as early as 700 million years after the formation of the Earth. That early in the history of the solar system, the Earth was quite inhospitable, however, and a toxic place for life forms remotely like the ones we are most familiar with now: a lot of terraforming followed over the subsequent billions of years.

The first phase of life on Earth was a period of single-celled, microbial life. That life persisted for 1 billion to 1.5 billion years before it became abundant and diverse enough for its combined photosynthesis to substantially change the chemical composition of the atmosphere: free oxygen, which had until then been quickly consumed in chemical reactions, if indeed it existed in any significant amounts at all, for the first time remained as a substantial chemical compound in the atmosphere after the Earth celebrated its two-billionth birthday. This was a fundamental shift toward the terraforming of the early Earth into an Earth that resembles the one we now see around us: oxygen replaced methane and carbon dioxide as the most abundant gas in the atmosphere after the always plentiful nitrogen.

Living organisms continued to terraform the Earth as they evolved into varieties that grew abundant enough to change not only the atmosphere but also other aspects of the environment in which they lived. They altered the appearance of the landscapes and collectively modified the global climate. It took another 2 billion years after the shift toward an oxygen-rich atmosphere, however, before multicellular organisms, either plant or animal, appeared. Humans, if we grant them a history of 200,000 years starting from the oldest remains that really look like us, have been on this Earth for only 1 part in 23,000 of its history. Our history as a civilized species, starting some 5000 years ago with the Bronze Age, covers approximately 1 part in 800,000 of the history of life on Earth. There is no doubt that all the organisms of the Earth's past combined, no matter how small they may have been, have had plenty of time to change the conditions on our home planet.

The vast number of organisms living on the Earth, including humans, use and re-use the available chemicals of life in a network of cycles that all together maintain the Earth's biosphere. As we saw in Chapter 6, there are four chemicals in those cycles that are most abundant in life: hydrogen, carbon, nitrogen, and oxygen. Together, these four elements add up to over 96% of the mass of the human body. All life on Earth relies on those elements, in combination with some 17 more that are far less abundant in our bodies. How much of these omnipresent four elements is available for the biochemical reaction chains in the life-bearing domains of the Earth is strongly influenced by the existence of that life: life in the present conditions of the biosphere exists as much because it evolved to adapt to its environment, as because it has shaped that environment by its very presence. Life and its environment mutually adjust to the changes in each other, as long as those changes are slow enough for evolutionary adjustments to occur. The life-sustaining web of chemical cycles not only spans all living organisms, however: it also extends well beyond the diversity of the biosphere, connecting into geological cycles that have existed since the birth of the planet.

The birth of the Earth was completed with the final cataclysmic collisions of proto-planetary fragments that left the nascent planet a glowing ball of magma with an atmosphere of vaporized rock. The heat of the Earth's hellish formation subsided within some millions of years, but before life pervaded the Earth's continents, oceans, and atmosphere, the planet went through a series of metamorphoses, fundamentally changing its appearance with each step. Where we are used to a blue–white ball with abundant green landmasses, the earliest Earth was a lifeless sphere glowing with cooling magma and with frequent volcanic eruptions spewing poisonous gases and Sun-obscuring dust into a super-hot atmosphere, into which even rocks would evaporate. Once the planet's surface cooled sufficiently for water vapor to condense into oceans, seas, and lakes, the atmosphere was presumably composed of mostly carbon dioxide. The pressure of that early atmosphere was likely 100 times that of today and equivalent to a pressure that now exists at depths of 1 kilometer in the oceans. The greenhouse effect of all that carbon dioxide kept the temperature hovering above 200 degrees centigrade (400 degrees Fahrenheit). At that temperature, only the high pressure of the dense atmosphere kept the oceans from boiling. In that heat, today's proteins would not be able to exist in the complex, folded state in which they guide chemical reactions that are essential to terrestrial life.

Over time, the bulk of the carbon dioxide that existed in the early atmosphere would be removed by a geological process that continues today. Whenever carbon dioxide comes into contact with silicates in rocks and with water, it is very gradually taken out of the atmosphere. First, carbon dioxide gas dissolves in water to form carbonic acid, which can be precipitated as an acidic rain. When this rain comes into contact with silicate materials in rocks, a process occurs that is called chemical weathering. Among other things, this weathering yields clays as well as calcium and bicarbonate ions. These ions combine to form calcium carbonate, which is what makes limestone and chalk. On the young Earth, this weathering process would remove most of the carbon dioxide from the atmosphere in 10 million to 100 million years, to lock it up in limestone and chalk formations. On geological time scales, ocean floors sink and dissolve into the Earth's interior. Together, these processes remove carbon dioxide from the atmosphere by transporting it deep into the Earth for a very, very long period of storage.

By the time the Earth was some tens of millions of years past its birth phase, the initial water vapor had condensed into oceans, and enormous masses of carbon dioxide had been removed from the atmosphere by weathering. Thereby the concentrations of two of the main greenhouse gases had been greatly reduced, and this caused the surface of the Earth to cool down further. It still remains to be determined just how cool the Earth actually became. Some hypothesize that it was far below freezing everywhere for a long time, with hundreds of meters of ice covering the entire surface, even the equatorial regions. One reason for such a cool early Earth is that the Sun—which has been very gradually brightening since its birth—would have been 30% less bright when it was young than it is today. Alternative hypotheses to this "snowball Earth" are a warmer Earth with relatively thin ice ("only" a meter or yard thick on average, and not covering all of the oceans) or even a rather hot Earth. There is evidence for each of these scenarios and scientific reasoning behind each of them. It is even possible that the Earth cycled irregularly through two or more different states. We do not know enough about global climate systems in the decidedly unterraformed conditions that reigned on the very young Earth, orbiting a substantially fainter Sun, to know for certain what the Earth's climate would have been like, and how it evolved over millions of years.

From the point of view of life on Earth, this uncertainty about the temperatures on the juvenile planet is not all that important. Some 700 million years after the formation of the Earth, the Moon, and the other planets, something caused a great upsurge in the number of cataclysmic collisions between interplanetary bodies. This "late heavy bombardment" by objects, many of which were at least 10 kilometers in size, created craters on the Earth with diameters in excess of 1000 kilometers. This bombardment is described as "late" because it occurred well after the main phase of formation of the planets in the solar system, although it remains to be determined whether it was relatively sudden in geo-logical terms or whether it was a bombardment spread out over a few hundred million years. Regardless of its duration, it appears to have oc-curred in much of the solar system, and was possibly related to a gravi-tational tug of war between the young heavy planets in the outer solar system acting on remnant rocks orbiting within it. Many of the enor-mous craters that were caused by the impacts with sizable bodies remain visible on the Moon, for one. Geological activity, weather, and life to-gether have erased all that evidence from the Earth. In fact, these com-bined processes are so efficient in reshaping rocks and entire landscapes that the oldest surviving rocks on the Earth, found in only a few places in Australia, Africa, and Canada, date back to just about that period.

The larger of the many meteorite impacts in the era of the late heavy bombardment may have been sufficiently severe to vaporize all of the oceans, if not even to vaporize a large amount of rock. Such gigantic impacts would have raised the atmospheric temperatures to many hun-dreds of degrees. Cooling in between major impacts might have taken several hundred years, gradually reducing the global temperature again to possibly below freezing if the "snowball Earth" hypothesis is correct. An intermediate view of the Earth during this period of bombardment is that the ice was thick at middle and high latitudes, that the equatorial seas were partly unfrozen or thinly covered with ice, and that multi-tudes of meteorite impacts could unfreeze the globe for a while until the temperatures settled back to their uncomfortably low values.

However long it may have lasted, it is by the end of the late heavy bom-bardment, some 700 million years into the Earth's history, that we can start looking at a few of the elemental cycles still in action today, but prior to the complicating actions of life forms that added their own cycles into the mix. The two cycles that have existed on the Earth for the longest time are the water cycle (in which much of the hydrogen and oxygen are being processed) and the carbon cycle. The surface–atmospheric water cycle, also known as the hydrological cycle, is the one that we best rec-ognize from daily experience: water evaporates from open waters, be-comes part of the atmosphere, where it often forms clouds, precipitates from the atmosphere in one of its various forms, and flows back to large bodies of water through rivers or underground pathways. In the early days of the Earth, living organisms did not participate in this cycle, but nowadays uptake by, and respiration from, vegetation are also important links within the hydrological cycle.

These days, over 97% of all water on the planet's surface layers is contained in the oceans as salt water, with another approximately 2% in
glaciers and ice sheets, where it is stored as fresh water. A mere 0.3% of
all water, fresh and salt, is contained in lakes, rivers, streams, and ground
water. All of that came down at some time as precipitation in the form
of rain, snow, or hail. Much of this falls into the oceans, of course, where
it is of little direct use to humans. Mankind utilizes somewhat in excess
of one-tenth of all surface runoff, on average, so that we are involved in
only a very small fraction of the overall water cycle. Locally, the fraction
of surface runoff used by humans may be much larger, of course, particularly in and near desert areas. Or it may be smaller, as it commonly
is in the wet tropics.

Water moves from a very large reservoir (the oceans and the seas)
into several small intermediary reservoirs, namely the atmosphere
and surface waters. From these smaller reservoirs, it cycles back into
the oceans and seas. Water spends a much longer time in the larger
reservoir than in the smaller ones. The average time that water molecules spend in the oceans and the seas is close to 3000 years. In contrast, they spend only just over a week in the atmosphere before they
precipitate out again, then a little over two weeks in streams and rivers
before once more reaching the oceans and seas. Ground water is a relatively small reservoir within this cycle, but water flows through this
very slowly: on average, ground water spends three centuries underground, although shallow ground water may dwell there merely for a
few years, while in deep reservoirs it may reside for tens of thousands
of years.

These estimations tell us that, on average, rain contains water that
has been in the oceans and seas since the Great Pyramid was built in
Egypt, while water from very deep wells may have been in the ground
since before the French and American Revolutions but in some cases
longer than the oldest cities on Earth have existed. These globally averaged numbers depend on the local conditions and the shifting climates:
some old groundwater is currently no longer being replenished, for
example, as relatively new desert areas have formed. One example of
this is the Sahara. Some 5500 years ago this vast segment of northern
Africa was mostly a savanna, as now exists in Kenya, Tanzania, and elsewhere on that continent. The region of the modern-day Sahara then
supported numerous herds of large animals, but nowadays it receives
no substantial rains, so that its deep groundwater, which percolated into

the subterranean rocks thousands of years ago, is not being replenished as fast as it is being lost.

There is another, very much slower cycle in which water is involved, which largely parallels the carbon cycle: water cycles deep into the Earth, trapped in rock and sediment, in geological transport mechanisms. As is true for much of the hydrological cycle, the carbon cycle was already partially in progress on the Earth before living organisms existed. That oldest segment of the overall carbon cycle, which does not involve living organisms, involves continental drift and volcanism and is lengthy and slow. Its overall complexity meant that understanding of its workings took a very long time to develop. In this cycle, carbon dioxide is removed from the atmosphere through the chemical weathering of rocks, which, as described earlier in this chapter, was instrumental in removing the vast amounts of carbon dioxide from the young Earth's atmosphere. Weathering is a slow process that would take millions of years to remove all atmospheric carbon dioxide. If there were no way to replenish the carbon dioxide that is being lost by this process, however, the atmospheric content of this gas would drop below the levels needed for plants and trees to survive. In the past two centuries, mankind is one of the significant sources of new carbon dioxide in the atmosphere because of the burning of coal, gas, and oil. But for something as old as the Earth, mankind's fossil-burning presence is very, very recent: for billions of years, the Earth's atmosphere has had another source of carbon dioxide, one that has its origin in geological processes.

Geology is a science that started in earnest only in the latter half of the seventeenth century. One of the drivers of that development was the existence of fossils. Fossils presented three major problems in the eyes of researchers from centuries past. The first problem was that many fossils had forms that matched no living animal or plant. Fossilized bones, for example, of various large dinosaurs or the large early mammals such as the mastodon were known in ancient China and Greece, and were viewed as remains of dragons, giants, and other mythical creatures. The absence of any living animals to which those fossilized ones could be matched was considered a problem from a religious point of view: why would any god create animals and then exterminate them, and why had no one mentioned them if, as many believed, the world was created only a few thousand years earlier, with most of that history covered in biblical texts? Were these missing animals and plants perhaps lost in a great flood of which the Bible tells?

The second problem of the fossils was that they were found deep in-side rock formations. This was clear to any architect of major buildings. The pyramids of Giza, just outside Cairo in Egypt, were once covered with limestone (later removed by Romans for their own use) in which many fossilized seashells were seen. The limestone quarries of Europe, which supplied building materials for many of the numerous medieval churches, were also full of fossils, even though these quarries cut into the sides of mountains to expose rocks from far below the Earth's sur-face. How could animals be buried so deep inside rocks in a world that was supposedly formed before the animals were added to it in a later phase of biblical creation? In those days, it was considered possible by some that fossils were generated spontaneously inside the rocks, as the rocks supposedly mimicked the world above.

Among those pondering these issues was Robert Hooke (1635–1703), an English scientist. In 1665, Hooke published a book about observa-tions that he had made with a microscope (an invention made in 1590 in the Netherlands, possibly by the same Hans Lippershey who some years later claimed also to have invented the telescope). Among the de-scriptions of living tissues, Hooke recorded his observations of wood, in which he saw—as in all living organisms—the fundamental structure of the cell (a term that he proposed). He also applied his microscope to fossilized wood and noted that it showed the same geometrical patterns as wood from living trees. These and other observations led him to con-clude that fossils had once been living organisms and that exposure to minerals had somehow transformed their structures into rock.

A year after Hooke's book was published, a Danish priest, Nicolas Steno, was working in the courts of the Medici in Florence. He was brought the head of a great white shark, a very unusual catch in the Mediterranean Sea, for study. A very observant naturalist, he noticed the similarity of the shark's teeth to fossils that he had seen high in the mountain ranges of Italy. Not only had he seen shark's teeth in the mountain rocks, but also fossil shells and the fossilized remains of other sea creatures. His inquisitive mind led him to tackle this issue, the third major problem with fossils: sea animals had somehow ended up as fos-sils high in mountain ranges. He did not accept the hypothesis of the day that these fossils grew spontaneously inside rocks, that the sea levels had been high enough to cover mountains (perhaps during Noah's flood), or that these fossils fell from the Moon or were planted there upon the creation of the Earth. Steno meticulously observed the Earth's makeup, and he was the first to propose the fundamentals of modern-day strati-graphic geology: layers of the Earth would form in a temporal sequence, with the oldest deepest in the Earth; they would form horizontally even if later they somehow became tilted; the layers would form over large areas, so that patterns would match from one location to another over some distance; and if there were breaks in these patterns then these must have occurred well after the original sedimentation.

These concepts were incorporated over a century later into James Hutton's theories of the formation of rocks, which included the idea that the ocean floors and landmasses were subject to uplifts and sub-mersions. In his words (from 1785, later printed in his book *Theory of the Earth; or an Investigation of the Laws observable in the Composition, Dissolution, and Restoration of Land upon the Globe*):

> Hence we are led to conclude that the greater part of our land, if not the whole, had been produced by operations natural to this globe; but that in order to make this land a permanent body, resisting the operations of the waters, two things had been required; 1st, The consolidation of masses formed by collections of loose or incoherent materials; 2ndly, The elevation of those consolidated masses from the bottom of the sea, the place where they were collected, to the stations in which they now remain above the level of the ocean.

Among his observations, he would eventually describe granite and vol-canic rock in fractured and bent sedimentary rocks. He concluded that stratified sediments had somehow reached depths where molten rock could mix with it. He concluded that rocks underwent the processes of deposition, uplift, deformation, and erosion many times over in geo-logical cycles of formation and destruction of entire landscapes.

Another century later, astronomer and meteorologist Alfred We-gener, when working in Marburg in Germany, created a nearly com-prehensive description of the history of the continents. In 1911, he had been reading a description of fossils of plants and animals that were identical on the European/African and American sides of the Atlantic Ocean. Further study revealed more such connections between con-tinents across oceans: mountain ranges lined up and stratifications of rock formations were in places identical on opposite sides of the water. The remarkably good side-by-side fit of some continents such as Africa and South America, when compared on model globes, had been noted before by geographers and others. Some, including Alexander von

Humboldt, who had died in 1859, had even remarked that these continents might once have been positioned close together. At the beginning of the twentieth century, however, the consensus among geologists was that the fossil similarities between continents reflected the past existence of land bridges, which had somehow sunk into the oceans, and that other geological and biological similarities across the oceans were simply there by chance.

When Wegener uncovered other facts to support his developing ideas, he proposed to the scientific community that lands do not simply move up and down, but that continents move sideways, drifting around across the globe. Only by postulating such colossal movement could he explain how fossils of tropical ferns could be found on the Spitsbergen archipelago at the confluence of the Arctic Ocean, the Greenland Sea, the Barents Sea, and the Norwegian Sea—none of them in any way reminiscent of tropical waters. His overall concept turned out to be correct, but because the mechanism he proposed presented difficulties to geophysicists at the time, his ideas were greeted with hostile criticism.

It would take another half century, until the late 1960s (over three decades after Wegener's death in a blizzard during an expedition to Greenland), before the mechanisms of continental drift were sufficiently well understood for it all to make sense. By then, scientists understood that the continents were not plowing through the ocean floors, but that the continents were moving with the surrounding ocean floors. The segments of the Earth's surface move as rigid plates at speeds of merely centimeters (or a fraction of an inch) per year; the solid plates float above the hot, liquid magma underneath, being as thin, in relative terms, as the skin of an apple compared with the whole. The topmost layer of non-solid rock on which the plates float is a soft, viscous domain of the Earth's mantle, known as the asthenosphere, from the Greek for "weak domain". It is in this domain that the slow magma that flows underneath couples to the solid continental plates above, through friction, which forces the plates to move about. Where a plate is fractured into two pieces that move away from each other, magma fills in the gap—generally in deep oceanic volcanic mountain ranges—cooling and solidifying to form new segments of the plates.

One of the final pieces of evidence that caused the idea of continental drift to be accepted as a well-proven theory came from the Earth's magnetism. By the early 1960s the concept of ocean-floor spreading was being worked on, when Fred Vine, his thesis advisor Drummond

Matthews (both British marine geologists), and their Canadian colleague Lawrence Morley found a way to demonstrate it beyond doubt. When magma rises into the gaps left by spreading ocean floors, it cools and solidifies. This basalt comprises a mix of minerals, one of which is magnetite. Magnetite, an iron oxide, is—as its name suggests—a magnetic substance: when heated to above about 585 degrees centigrade (or 1085 degrees Fahrenheit) it loses its magnetic field, but when it is cooled again to below that temperature, its structure takes on the strength and direction of the magnetic field that exists in its surroundings at that time. For cooling basalt, that means that the direction of the Earth's magnetic field becomes locked into the rock. Because the Earth's magnetic field changes fairly rapidly compared with the geological time scales required to move deep oceanic crust over considerable distances, the pattern of the changing magnetism is frozen into the rocks as the cooled basalt formations spread apart on both sides of the zone of their formation. This results in banded patterns in the rock's magnetism, laid out like symmetrical bar codes on either side of the rift. The realization by Vine, Matthews, and Morley of how these patterns formed was the final argument proving continental drift.

The Earth's globe is a finite object, so if new seafloors form where plates move apart, somewhere else plates must be moving into each other. Where plates collide, one plate must yield to another by moving underneath it, while the other is generally crumpled and pushed upward in a process that forms mountain ranges. At such places where two plates are pushed together, generally deep trenches are found next to mountains; an example of this is found on the Pacific coast of Chile, where the high Andes mountains lie next to the deep Atacama Trench. The rise of the mountains on a continental plate generally occurs in intermittent spurts as the adjacent oceanic plate slides underneath them. We know these growth spurts of mountains as earth tremors or earthquakes.

The rise of the ground in one such earthquake was described by Charles Darwin after he experienced an unusually strong event in Chile in 1835. In a letter home, he wrote:

> We are now on our road from Concepcion. The papers will have told you about the great Earthquake of the 20th of February. I suppose it certainly is the worst ever experienced in Chili. It is no use attempting to describe the ruins—it is the most awful spectacle I ever beheld. The town of Concepcion is now nothing more than piles and lines of bricks, tiles

and timbers—it is absolutely true there is not one house left habitable; some little hovels built of sticks and reeds in the outskirts of the town have not been shaken down and these now are hired by the richest people. The force of the shock must have been immense, the ground is traversed by rents, the solid rocks are shivered, solid buttresses 6–10 feet thick are broken into fragments like so much biscuit.[...] I am very glad we happened to call at Concepcion so shortly afterwards: it is one of the three most interesting spectacles I have beheld since leaving England—A Fuegian Savage—Tropical Vegetation—and the ruins of Concepcion. It is indeed most wonderful to witness such desolation produced in three minutes of time.

Darwin not only noted the devastation in that letter, but also documented the rise of the land in his diaries, which he later used as the foundation for his *Voyage of the Beagle*. His entry for March 4, 1835, reads:

The most remarkable effect of this earthquake was the permanent elevation of the land, it would probably be far more correct to speak of it as the cause. There can be no doubt that the land round the Bay of Concepcion was upraised two or three feet; but it deserves notice, that owing to the wave having obliterated the old lines of tidal action on the sloping sandy shores, I could discover no evidence of this fact, except in the united testimony of the inhabitants, that one little rocky shoal, now exposed, was formerly covered with water. At the island of S. Maria (about thirty miles distant) the elevation was greater; on one part, Captain Fitz Roy found beds of putrid mussel-shells *still adhering to the rocks*, ten feet above high-water mark: the inhabitants had formerly dived at lower-water spring-tides for these shells. The elevation of this province is particularly interesting, from its having been the theatre of several other violent earthquakes, and from the vast numbers of sea-shells scattered over the land, up to a height of certainly 600, and I believe, of 1000 feet. At Valparaiso, as I have remarked, similar shells are found at the height of 1300 feet: it is hardly possible to doubt that this great elevation has been effected by successive small uprisings, such as that which accompanied or caused the earthquake of this year, and likewise by an insensibly slow rise, which is certainly in progress on some parts of this coast.

Darwin had directly witnessed the phenomenon of uplift, one of the key processes in the geological carbon cycle.

Knowing of continental drift, we can now outline those aspects of the carbon cycle that occur whether living organisms exist on the planet or not. The weathering of rocks captures carbon dioxide from the atmosphere and thereby makes it part of the lithosphere, the outermost rocky shell of the planet, in limestone deposits. Tectonic plate collisions cause some of these deposits to sink into the Earth's interior as one plate moves underneath another. Once deep in the Earth, the rock formations dissolve into the continuum of magma. New parts of the lithosphere form in oceanic ridges where plates separate, as volcanic activity occurs. This also happens elsewhere, where colliding plates cause mountain ranges to form, and in certain hot spots where the Earth's mantle is unusually thin, causing isolated volcanoes to form. Eruptions of these volcanoes release carbon dioxide, along with other gases, into the atmosphere. Without living organisms, the amount of carbon dioxide in the atmosphere is thus a balance between the processes of the weathering and the movement of continental plates into the Earth's deep interior, which remove carbon dioxide from the surface, and volcanic activity, which reintroduces it again. The characteristic globally averaged time for this part of the overall carbon cycle is approximately 150 million years.

For the present-day Earth, the geochemical cycle just described amounts to about half of the overall carbon cycle. Given the ubiquitous presence of living organisms on the Earth, the overall carbon cycle is composed of a hierarchy of stacked cycles, of which the shortest take less than a year while the longest takes over 100 million years. These cycles are all interlinked. We can think of this in the same way that our lives are composed of hierarchically stacked cycles: the quick cycles of the heart beat and respiration, within the daily cycle of our work, within the social cycle of the week, within the annual cycle of the seasons and the holidays, all nested within the cycle of life from birth, to aging, and death.

The bulk of the carbon that now exists within our bodies has experienced an analogous hierarchy of cycles before it came to be part of us: carbon spends approximately 3 years in the atmosphere, 5 years in plants, 30 years in soils, 300 years in the oceans, and 150 million years in the geochemical cycle. The longest cycle also holds the largest reservoir: 99.95% of all carbon in the lithosphere, biosphere, and atmosphere combined resides in carbonates, that is, in rock. In that overall cycle, only about 40 carbon atoms in a million reside in fossil fuels. The fossil fuel reservoir is more than five times larger than the even smaller reservoir of the atmosphere, however, so that the atmospheric carbon dioxide content would go up dramatically if we burned most or all of the fossil fuels. The geochemical cycle is far slower at removing carbon dioxide (i.e. through weathering and geological movements) than the rate at which humans are putting it into the atmosphere. Consequently, the carbon dioxide content of the atmosphere is rising: the increased carbon dioxide content does somewhat accelerate the action of weathering subject to increasingly acidic rain, but mankind's addition to atmospheric carbon dioxide through the burning of fossil fuels exceeds by far what even that increased weathering removes.

The duration of the geochemical carbon cycle, very roughly 150 million years, is short enough that it has cycled the surface carbon through the Earth's interior several times since about 575 million years ago, when complex life forms first appeared. Such multicelled organisms with differentiated cell functions were new then, in a world that previously hosted only single-celled organisms. Some 540 million years ago, life diversified tremendously in what is known as the Cambrian explosion, which lasted some 50 million years. Sometime after that, the first land plants emerged, and by 215 million years ago dinosaurs roamed the Earth and some larger animals developed flight. A major meteorite impact some 65 million years ago appears to have been the last blow for the dinosaurs (except for the smaller kinds that later evolved into modern-day birds). That time of dinosaur extinction coincided with the beginning of the era of the mammals and of the flowering plants.

Since the first mammals appeared, the continents have continued to move around. The vast old supercontinent of Gondwanaland has fragmented. Antarctica has drifted to the south pole to become a small-scale model of the early "snowball Earth", and the Indo-European and American continents took much of their present shape in the mammalian era. At the very end of all this geological and biospheric activity, something somewhat akin to the modern human came into being some 2.5 million years ago. Since then, not a great deal has happened in geological terms: the continents continue to move around, and in some of their collisions young mountain ranges continue to be raised (as is happening in the Himalayas and in the Swiss Alps), but overall the distribution of the lands has been as it ever was seen by mankind.

The carbon that is now in our bodies has most likely been cycled through the Earth's interior just once since the first dinosaurs appeared 215 million years ago. The hydrological surface–atmosphere cycle is very much faster than any geological cycle: ocean waters have been cycled through rain and snow an estimated 70,000 times in that same time frame. Rocks and magma in the Earth's interior harbor water too, however, which participates in a much longer cycle. This locked-up amount may be comparable (with a large margin of uncertainty) to the volume at the surface of our planet. The cycle time to bring that from the surface into the Earth and back is likely similar to the geological carbon cycle time: water vapor introduced into the atmosphere by volcanic eruptions puts some very old water back into the hydrological cycle and thereby, ultimately, into our bodies.

The last of the biogeochemical cycles affecting the dominant four elements in our bodies is that of nitrogen. The bulk of the nitrogen on the planet, 83%, is contained in chemical compounds in rocks and sediments. Nitrogen in that reservoir is locked up so effectively that it is not of much use to living organisms. The atmosphere is the next largest nitrogen reservoir: the air, which is 80% molecular nitrogen by mass, contains 17% of the total nitrogen content of the planet. Despite the vast amounts of nitrogen in the atmosphere, nitrogen that can be used in the chemical cycles of life forms is in remarkably limited supply. This is because atmospheric nitrogen molecules contain two nitrogen atoms that are particularly greedy with respect to electrons, so that these two atoms are held together by an unusually strong electrical bond. The strength of that bond makes molecular nitrogen very hard to use chemically, be it by plants, by animals, or in human industrial processes, because it takes a lot of energy to break. Once broken, however, a reforming bond releases an equivalent large amount of energy. The potential to release a lot of chemically stored energy from nitrogen compounds made the earliest human uses of nitrogen as much for military-industrial applications (such as in the use of saltpeter for gunpowder) as for agricultural applications (where it serves as fertilizer).

For living organisms, usable nitrogen in their environment derives mostly from ammonium, nitrate, and nitrite. These forms are referred to as "fixed nitrogen", in which nitrogen is bound to other substances (to hydrogen in ammonium, and to oxygen in nitrate and nitrite), to contrast it to the free gaseous nitrogen in the atmosphere. Fixed nitrogen is produced in three fundamentally distinct processes, only two of which occur naturally. The first and oldest of these natural processes is lightning. In the present-day world, some 6% of fixed nitrogen is created by reactions that start in lightning storms: lightning causes nitrogen and oxygen to combine into nitric oxide and nitrogen dioxide, which then combine with water vapor into nitric acid, which is subsequently dissolved in rain and thus transported into the soil. Lightning would have been active also on the pre-life Earth, and may have been critical in creating the compounds on which life rests: nucleic acids and amino acids. In the early Earth's atmosphere, there would have been no free oxygen, but an experiment conducted by Stanley Miller and Harold Urey in 1952 demonstrated that other reaction pathways would have existed that were enabled by lightning: driving electrical sparks through a mixture of water vapor and nitrogen gas, mixed with methane and ammonia in a closed container in their laboratory as a tiny recreation of the natural setting, they showed that complex molecules, including multiple amino acids, formed without further intervention. Experiments with other chemicals thought to exist in the Earth's early atmosphere, such as carbon dioxide and hydrogen sulfide, yielded comparable results.

The second of the nitrogen fixing processes, creating approximately 60% of all the ammonia used by living organisms, occurs by bacteria and archaea on land or by cyanobacteria, also known as blue-green algae, in the oceans. Many bacteria have an enzyme, called nitrogenase, that converts molecular nitrogen into ammonium, and then these bacteria and others subsequently process that captured nitrogen into nitrate. Plants generally take dissolved ammonium or nitrate from the soil. Plants in the legume family, which includes all beans, have a symbiotic relationship with nitrogen-fixing bacteria that live in nodules in their root systems, where they exchange ammonia for carbohydrates from the plants. Land animals and fungi obtain all their nitrogen from the consumption of plant materials (mostly in the form of amino acids), either directly or via the digestion of compounds from animals that ingested plant materials prior to being eaten.

The third major pathway for nitrogen fixation is through human industrial activity: over 30% of all nitrogen fixation on Earth, and over 40% of all nitrogen fixation outside of the oceans. At high pressures and temperatures, and with the help of catalysts, nitrogen-containing fertilizers and explosives can be produced. Fertilizers provide a major boost to the yield of crops to feed mankind and keep soils productive by reintroducing nutrients that plants take out of the soil when growing. It is estimated that about half the food production in the world is presently enabled by artificial fertilization. Prior to artificial fertilizers, agriculture had but few options to counter the exhaustion of nutrients from the soil: apply natural fertilizers (indirectly using animals to concentrate nitrates by using their excrement in the form of manure and guano), move on to new land (slash-and-burn agriculture), await natural floods to bring in new nutrients (as practiced of old around the major rivers, but now increasingly rare as hydroelectric dams are built), use crop rotation (because some crops—particularly the legume family—can fix nitrogen through their symbiotic relationship with nitrogen-fixing bacteria), or use blue-green algae in crops that grow in wet environments (such as rice that grows in paddies that are flooded for part of the growth period).

The nitrogen cycle through the vast reservoir of nitrogen in the atmosphere is quite slow: nitrogen spends on average 14 million years in the atmosphere before being used by plants or animals. Because humanoids have existed for no more than 2.5 million years, this means that most of the atmospheric nitrogen has never been inside a human being (provided we do not count time spent in the lungs as being really inside). It also means that it has been in plants or animals only about 35 times since multicellular life developed on the planet. Nitrogen spends an average of three years in the land plant reservoir before cycling into another reservoir. Humans and other animals get all their nitrogen by eating plants, plant eaters, or fungi.

The interlinked cycles of water, nitrogen, oxygen, carbon, and other elements (notably phosphorus and sulfur) that connect the organic and inorganic spheres, or the living and the lifeless worlds, continue to terraform planet Earth. The relative stability of life and the climates in which it occurs is largely a direct consequence of the stabilizing feedback loops in the overall system. If, for example, volcanic activity were to increase the carbon dioxide content of the atmosphere, the greenhouse effect would raise global temperatures, and the resultant increased plant growth and more efficient chemical weathering would increase to ease the carbon dioxide level back down again. Such periods of enhanced volcanic activity do occasionally occur; one in the latter half of the seventeenth century is argued to have contributed to what is referred to as the Little Ice Age, during which many crops in Europe failed and winters were unusually severe. In that case, the effects of volcanic dust blocking light from the Sun were possibly more important, however, than the increased carbon dioxide content of the atmosphere, and caused an effect opposite to the one we might have expected.

Another type of biosphere–geochemical climate feedback occurs because of the gravitational pulls of the Sun, the Moon, and the major planets (mostly Jupiter and Saturn). The pulls of the Sun and the Moon cause a very slow wobble in the tilt of the Earth's rotation axis relative to the Sun (with a period of 41,000 years). That tilt is the primary reason for the climate to exhibit seasons, but the wobble in the tilt will slowly affect the seasons, and in turn cause the global climate to adjust. The gravitational pulls of the major planets force slight changes in the orbit of the Earth around the Sun, which causes the Sun to appear with slightly modified brightness in the Earth's skies (on time scales of hundreds of thousands of years). That, too, modulates climate.

That the Earth's orbit and tilt were not always the same was first fully worked out by the Serbian geophysicist Milutin Milankovitch. He did this in part after being imprisoned while on honeymoon as a foreign national in the Austro-Hungarian Empire, at the start of World War I. (His being Serbian was a particular problem because the war purportedly started upon the assassination of both the heir to the empire and his wife by Bosnian-Serb activists in Sarajevo.) Milankovitch's wife managed to mediate his transition from prison to house arrest in Budapest, Hungary, where he was allowed to continue his work. He published it in 1916, in the middle of the war. The result of his work is that the evolution of the various gravitational effects on the Earth's orbit and tilt are now known as the Milankovitch cycles.

How does the Earth's climate respond to these changes in the strength and distribution of sunlight over the globe? We can expect that any climate change will cause changes in plant growth and in the weathering of carbon dioxide. If the global climate becomes warmer, plants likely grow faster over much of the globe, while chemical weathering also increases. That removes carbon dioxide more effectively from the atmosphere, which counters the increase in global temperature. Exactly the opposite happens if the global climate becomes cooler. Thus, the climate effects of the gravitational pulls of the Sun, Moon, and planets are weakened compared with what would happen without life and weathering, at least on time scales of tens of thousands of years. This does not completely work, however: the Milankovitch cycles are recognizable in the long-term climate records of the periodically occurring ice ages that blanketed much of the Earth in massive glaciers. This must mean that the stabilizing effect of living organisms on climate variability has its limits.

To explore more quantitatively how the terraforming effects of living organisms could—within limits—act to stabilize their own ecosystem, including the global climate, James Lovelock and Andrew Watson devised a simple computer experiment. They called their computer model "Daisyworld". Even with the limited computer capabilities available at the time (1983), they could run an insightful simulation of a virtual world that contained two types of digital "plants": black daisies and white daisies, which reflect sunlight differently and have somewhat different growth rates as a function of temperature. The early computer experiment revealed how two different populations of virtual vegetation could maintain a relatively beneficial environment for both species even if subject to substantial changes in the model's incident sunlight. Later additions of other virtual species by different experimenters suggested at least that a virtual world could be stabilized further by the combined interactions between these species.

Modern computers allow vastly more complex numerical experiments, of course. These are beginning to uncover how the multitude of pathways that regulate greenhouse gases, climate, and life on land and in the oceans are actually coupled. But understanding the problem of climate in detail is an enormous challenge: not only are there many complicated chemical and physical processes involved, but mankind's response to climate change is even harder to capture in equations that computers can deal with. How do we predict how carbon dioxide output develops as cultures change, and as economies are subjected to changing sea levels or regulatory constraints? How does one capture societal conflicts when resources such as food and water become scarce in some places as rainfall or temperatures change, or as lumber resources for building materials become difficult to obtain in others? How does one capture in a computer code how long society may almost entirely ignore the scientists' findings that climate is changing without adequate responses to reverse the process, or at least halt it? With so many unknowns, it is no surprise that there is no consensus on how far climate will shift subject to mankind's efforts to harvest the solar energy captured in ancient fossil fuels rather than harvesting solar energy directly or through the intermediaries of wind, waves, or hydroelectric energy. One issue with understanding climate change is that it happens relatively slowly on a human time scale, because the atmosphere, the oceans, and the polar ice fields are such vast reservoirs of energy and water. Consequently, observed changes appear small and slow. But once in motion, these effects are exceptionally hard to stop and reverse. It is like pushing a large ship or a heavy train: it takes much effort to get them moving, but once moving they have so much momentum that it is difficult to stop them again unless comparable efforts are expended.

Geological studies have shown ample evidence that there are limits to the stabilizing feedbacks that have kept the global climate fairly constant over the past thousands of years. After all, past ice ages of various strengths and durations show that major changes do occur in the global climate. The consequences of a substantial increase in carbon dioxide on a time scale that is much shorter than the geochemical cycle, such as mankind is now effecting by burning fossil fuels, are yet to be fully understood. The threat of climate chaos, the eventual magnitude of climate change, and the possibility of reversal of mankind's detrimental actions are being studied intensely. Lives may well depend on it.

Key Points: Chapter 7

- Microfossils of the earliest single-celled life forms date back over 2.5 billion years, but indirect evidence suggests that life may have emerged about 3.9 billion years ago, 700 million years after the formation of the Earth.

- Life on our planet evolved to adapt to its environment and has in turn shaped that environment (including climate). Mutual adjustments are possible, regulated through stabilizing feedback loops, as long as the changes are slow enough.

- Terraforming occurs by interconnected biochemical cycles that connect the organic and inorganic domains on our planet. The two oldest major chemical cycles are the water cycle and the carbon cycle, which connect us to the deep interior of the Earth. The nitrogen cycle is intimately connected with life through the plant nutrients nitrate and ammonia. All cycles are affected by the gravitational pull of the Sun, Moon, and planets.

- Climate change, both natural and induced by recent human activity, appears relatively small and slow on a human time scale because the atmosphere, the oceans, and the ice fields are vast reservoirs of energy and water. Once in motion, however, the effects can be profound and exceptionally hard to stop, let alone reverse.

• **Interdependent bio-geo-chemical cycles are linked through feedback loops that sustain life and influence our atmosphere, environment, and climate.**

In life, nothing is to be feared, everything is to be understood.

MARIE CURIE (1867–1934)

The ordinary novel would trace the history of the diamond-but I say, "Diamond, what! This is carbon." And my diamond may be coal or soot and my theme is carbon.

DANIEL HERBERT LAWRENCE (1885–1930),
in a letter to Edward Garnett (1914)

Infant Atoms

Most of the carbon on the Earth, and consequently in our bodies, has been here since the planet formed. Some 4.6 billion years ago, a huge cloud of gas and dust started to collapse under its own gravity. Within it, dust particles collided and stuck together to form increasingly larger clumps. Over time, these clumps kept falling onto each other, attracted by their mutual gravity, and conglomerated to eventually form full-sized planets. In the first tens of millions of years, collisions between the young planets and the remaining clumps of matter swarming between them would have been very frequent. Gradually, the larger bodies collected more and more of the material coming close to them, as their gravitational fields increased in strength with each collision and capture. The frequency of the collisions decreased as the remaining clumps were increasingly swept up by the pull of the planets, which cleared the planetary system of most of the initial debris. Eventually, the planets collected essentially all of the primordial fragments, locking the elements of life into their gravitational field for the remainder of the billions of years of their subsequent history. Since then, however, an ongoing cosmic shower has added a little more matter to the Earth every year. This ancient cosmic dust mixes with matter already there, and it is all used by living organisms.

The gravitational cleanup early in the history of the solar system (which we describe in more detail in Chapter 13) did not occur equally efficiently everywhere. Between the orbits of Mars and Jupiter still float a multitude of objects that we know as asteroids. Neither did the big cleanup reach into the outermost regions of our planetary system. To this day, there is an enormous reservoir of frozen objects out there, tens of thousands of times further from the Sun than the Earth is. Occasionally, some lumpy object in the vast collection of non-planetary bodies that orbit the Sun is pulled from its orbit by stars that slowly drift by our home in space at great distances. Those objects coming in from far away have always been in the cold, and are therefore composed mostly of gases frozen solid since before the Sun lit up. When these fluffy clumps of snow and dust come close to either the Sun or to a planet, they can be pulled apart by slight differences in gravity between the parts closer to and further from the big body that they are passing by. Alternatively, the heat of the Sun's light causes part of their structure to thaw and turn to gas, releasing other locked-in fragments in the process. All this causes dust, grains, pebbles, and rocks of a variety of sizes to populate the solar system, including near the orbit of the Earth. Over time, many such orbital perturbations of non-planetary bodies happen, which supplies a steady stream of particulate matter into all manner of trajectories throughout the space between the planets.

Nowadays, large objects collide with the Earth only infrequently. A very large crash, capable of affecting life on a large section of the globe, happens perhaps once every 100 million years. The last time such a big object hit the Earth altered the dominant types of life on the planet fundamentally: 65 million years ago a major impact onto Mexico's Yucatan peninsula is generally thought to have caused the large dinosaurs to become extinct, which in turn appears to have helped the diversity and number of mammals to increase.

Most of the matter falling onto the present-day Earth is not in the form of massive, heavy meteorites, however. The bulk of that matter is nowadays accumulated as dust with sizes typically much smaller than a grain of sand. Depending on the size and the trajectory of the impact, some dust may survive its descent through the Earth's atmosphere, while other particles are heated to evaporation by friction against the atmosphere as they come in at several kilometers or miles per second. The supply of new material from the distant reservoirs of dust in the solar system results in an ongoing shower of interplanetary debris, through which approximately 40,000 metric tons of material is added to the Earth every year, adding little by little to the total mass of the Earth. Even though that accretion rate may sound like a lot, if we take the annually captured meteorite mass over the entire Earth and multiply it by the lifetime of the Earth, it adds up to only 1 part in 30 million compared with the total mass of the planet. So, as far as the Earth is concerned, this dust collection over the billions of years since the early phases of its formation is negligible. We do not notice this dust shower among all other forms of dust in our atmosphere, but high-flying stratospheric aircraft have managed to scoop up some samples of this cosmic dust on its way down.

Even though this accretion is irrelevant to the planet as a whole, it makes a small contribution to all living organisms on the Earth: some of the material in our bodies, including the carbon on which we focus in this chapter, came from outer space some time during the history of the Earth. Plants and animals have been using the Earth's reservoir of carbon over and over again in countless generations of organisms, from the single-celled bacteria up to the largest living organisms, including the giant redwood tree and the blue whale. The variety of carbon-based chemicals that enables the functioning of every cell in every organism that exists or has ever existed is collectively known as organic matter. Some part of our food, and therefore of the material that makes our bodies, came to the Earth sometime between weeks to billions of years ago after having moved through the solar system as an asteroid or comet.

There is another component of the carbon in our bodies that has not been on the Earth for very long. Our knowledge of that component required the development of nuclear physics, which started not long after Henri Becquerel, a French scientist, discovered radioactivity, just before the beginning of the twentieth century. Scientists noticed that, curiously, some of the carbon could not have existed on Earth until quite recently. Eventually, they learned that it did not fall to Earth from elsewhere in the universe either. This realization started with the discovery that some carbon atoms are unstable and survive typically only for some thousands of years, rather than for billions of years as do the other carbon atoms. These unstable carbon atoms are radioactive and, over time, they transition into nitrogen atoms as they each eject one particle from their nuclei.

How can one element change into another? A chemical element is characterized by how many electrons orbit its atomic nucleus, which determines its chemical properties, and by the matching number of positively charged protons in that nucleus. Any process that changes the number of protons in an atomic nucleus eventually leads to a corresponding adjustment in the number of orbiting electrons and of the associated chemical properties of that atom. Changing the atomic nucleus is the only way in which one chemical element can be transformed into another.

The protons in the atomic nuclei are mixed with neutrons. Nuclei in the atoms of a particular chemical element have a fixed number of protons, but may have different numbers of neutrons. Nuclei with a given number of protons but with different numbers of neutrons are known as isotopes of that element. Carbon has three isotopes that occur naturally (i.e. outside of nuclear reactors). The most common isotope of carbon, ^{12}C, has six protons and six neutrons in its nucleus, hence the number 12 in its full physical and chemical designation. This particular mixture of protons and neutrons comprises close to 99% of all carbon atoms on Earth. That dominant isotope is not radioactive at all. The second most common form of carbon, ^{13}C, with six protons and seven neutrons in its core, is also not radioactive. That carbon isotope makes up a little over 1% of all carbon. It turns out that the radioactive carbon is by far the least common isotope, namely ^{14}C, with six protons and eight neutrons. That is the unstable form that can transform from carbon, with six protons, into nitrogen, with seven protons. It can do that by turning one of its neutrons into a positively charged proton while ejecting a so-called beta particle, better known as the negatively charged electron, from its nucleus. This type of transition, the ejection of a beta particle, is one of the different pathways of radioactive decay that exist in nature.

This least common form of carbon, also known as carbon-14, makes up only one part in a trillion of all carbon. In other words, there is only one carbon-14 atom among a million times a million atoms of carbon-12. Despite that small fraction, these carbon-14 atoms are well worth paying attention to: even with only one in a trillion carbon atoms being carbon-14, that still puts close to 1000 million times a million atoms of carbon-14 inside the human body. Dividing that number by the number of cells in the human body yields, on average, approximately a dozen carbon-14 atoms naturally present in each of our cells. Our body protects us from the effects of the natural radioactivity of carbon-14 very efficiently. Carbon-14 atoms are of interest, however, precisely because of their radioactivity, as that means that these must be infant atoms, very much younger than all the other carbon on Earth.

The minute amount of carbon-14 on Earth compared with the reservoirs of carbon-12 and carbon-13 has a remarkably history: carbon-14 atoms that now exist on Earth have typically existed as carbon for at most a few thousand years, rather than for at least 5 billion years as have the carbon-12 and 13 atoms. Put another way, the average carbon-14 atom has an age of less than one-millionth of that of carbon-12. It turns out that the story of carbon-14 ties us to dangerous radiation coming to us from outside the solar system, from deep within the Galaxy. That link is not direct, however: our atmosphere shields us from that radiation very well, but carbon-14 shows up as a side-effect of that very shielding.

The discovery of the linkage between radioactive carbon and radiation from the Galaxy started almost a century and a half ago. The last three decades of the nineteenth century and the first two of the twentieth century were times of rapid change for physicists. Fundamental discoveries about the nature of matter and physical phenomena were being made at a remarkable pace. In that period, physicists discovered, among other things, the most common building blocks of matter: electrons, protons, and neutrons. They also unveiled one of the most fundamental physical processes in the universe, namely the possibility that chemical elements could transform from one into another. For centuries, alchemists had been trying to achieve this by chemical means, but all their efforts were bound to be futile: the transformation of one element into another requires processes that cannot be chemically induced. At the start of the twentieth century, however, it was discovered that nature had other forces that could transmute elements. This scientific paradigm shift resulted in the picture of matter accepted today: all the chemical elements are made up of atoms with nuclei that are clusters of tightly packed protons and neutrons, with electrons orbiting relatively far away from these nuclei.

Our knowledge of the atom was assembled from very diverse pieces of evidence, key to which was the discovery of radioactivity. At the very end of the nineteenth century, the work of Henri Becquerel, Marie and Pierre Curie, Ernest Rutherford, and their colleagues had demonstrated that certain chemical elements emitted rays even when there was no heat or light around to cause that. These rays had to originate from within these substances themselves. Ernest Rutherford and his laboratory colleagues eventually demonstrated that radioactivity was the result of the transmutation of atoms from one chemical element into those of another by fragmentation, or fission, of the atom's nucleus. The products of radioactive decay, at first called Becquerel rays, are in fact signatures of the spontaneous splitting of atomic nuclei into small and large fragments. The smaller fragment created during radioactive decay is generally a free proton or neutron, an electron (still sometimes called a beta particle, for historical reasons) or its anti-particle, the positron, or a helium nucleus (often called an alpha particle, also for historical reasons). The other fragment resulting from radioactive decay is typically the nucleus of another element, much heavier than the smaller ejected fragment.

At first, the rays emanating from compounds of, for example, uranium, radium, or polonium were recorded using photographic plates. However, Marie Curie discovered that radioactivity could also be measured by an electrometer, a device that can store an electric charge and with which one can measure how rapidly that charge is lost. A very sensitive version of the electrometer had been invented by her husband, Pierre, and his brother, Jacques. She noticed that when a radioactive compound was placed near this electrometer, any electrical charge previously put on that device would leak away unexpectedly quickly, much faster than when the same device was placed far away from such compounds. Later, scientists understood that the "Becquerel rays" from the radioactive compounds would penetrate the walls of an electrometer and pass through the air within it. Some of these rays would even leave the electrometer again by traversing the opposite wall. While passing through the electrometer's container, the collisions of these rays with molecules of the air within it would cause electrons to be knocked from their orbits around the atomic nuclei. Normally, air is a very poor conductor of electricity. However, the presence, if only temporarily, of free positive ions and negative electrons within the air increases its conductivity enough to let the electrometer's charge leak away much more quickly.

The sensitive electrometer measured the very weak electrical current through the device as its charge leaked away through the effects of a radioactive substance placed within the device itself. The Curie quartz electroscope was another of their inventions, based on an effect discovered by the two brothers around 1880: the piezoelectric effect. They discovered that when stress is applied to certain crystals, they become electrically charged. When Gabriel Lippman predicted that the opposite should also occur, they had the tool to measure the weak currents induced by radioactivity: they learned how to balance the loss of charge within the electrometer by a current that was induced by putting weights on a quartz plate; this explains why the early measurements of radioactivity by the Curies and their colleagues were expressed in units of weight. In similar ways, the piezoelectric effect is nowadays used in a multitude of devices, ranging from the crystalline clocks that enable computers to function to the very fine movement needed to point and stabilize mirrors in the best astronomical telescopes.

As ever more sensitive instruments were developed to study radio-activity, the list of substances known to be radioactive grew. Carbon-14 was one of those. Shortly after World War II, scientists led by Willard Libby, working at the University of Chicago, measured how long carbon-14 atoms can expect to exist prior to decay into nitrogen. The lifetime that they first determined was close to the current best value: carbon-14 has a half-life of 5730 years. In other words, after 5730 years one in every two carbon-14 atoms will have transformed into a nitrogen-14 atom by a radioactive transition. After another 5730 years, only one in four survives. Then, with every half-life going by, there is only one in 8, one in 16, one in 32, and so on.

The half-life of carbon-14 fits into the Earth's lifetime almost 800,000 times. Even if the entire Earth had consisted of carbon-14 when it was formed, almost all of it would have vanished within 2 million years. Yet the Earth, over 2000 times older than that, still contains carbon-14. It may not be much, but it is more than enough to be thoroughly incompatible with carbon-14 having been on the Earth since the formation of the planetary system. Somehow, carbon-14 must continue to be added to the Earth's carbon reservoir. In fact, even if we allowed cabon-14 to originate from outside the Earth, such as in the shower of meteoric dust, it must have been created within a matter of thousands of years, somewhere within the universe close enough to reach us before it has decayed into nitrogen-14.

Physicists in the late nineteenth and early twentieth centuries realized that radioactivity was a property of multiple substances within the Earth. They knew that Becquerel rays associated with radioactivity could penetrate matter, but that it could not do so very far. The rays were absorbed fairly readily by solids and gases, including air: the more matter there is between a source and a measuring device, the more effective the shielding, and consequently the lower the intensity of the rays that make it to the measuring device from the source. When it was discovered that there was a weak background of Becquerel rays everywhere, one way to find out where those rays originated was therefore to measure how their intensity would change with increasing height above the Earth: if such rays were emanating from the Earth, then they should be increasingly absorbed by the ever thicker layers of air as a measuring device is raised higher and higher.

Theodor Wulf was one of the early experimenters trying to measure if that was the case. In 1910 he took an electrometer from the base of the

Eiffel tower to its top. The Eiffel tower was then the tallest building in the world, completed 21 years prior to Wulf's experiment, for the Paris world fair of 1889 that celebrated the centennial of the French Revolution. Wulf concluded that the radiation background that caused his electrometer to lose charge decreased somewhat when going from the base of the tower to the top, but that that change was far less than it should have been if all the radiation came from the Earth and traveled through absorbing air to the top of the tower. That measurement suggested, but could not unambiguously prove, that there was radiation coming from above the Eiffel tower, as well as from below. What would be needed to convince everyone that there was radiation coming from outside the Earth was the direct demonstration that radiation increases with height over what it was anywhere near the surface of the Earth. Just a few years after Wulf's original experiments, Victor Hess demonstrated just that. In a series of balloon flights he carried an electrometer up to 5300 meters (almost 20,000 feet). He found that the radiation was four times more intense at his highest altitude than at sea level. These rays had to come from outside the Earth, being absorbed by the atmosphere as they traveled into it toward sea level. Hess also flew in a balloon to perform that experiment during a solar eclipse, to rule out that these rays came directly from the Sun. In 1936, Hess received a Nobel Prize for his discovery. Some years later, Robert Millikan used the term "cosmic rays" to describe these extraterrestrial rays. That name stuck.

Cosmic rays are part of the naturally occurring low levels of radioactivity that the human body is exposed to all the time, with generally little lasting damage or health effects. We now know that they provide a window into what happens far, far away in the Galaxy. We eventually learned that their presence also informs us about the variability of the Sun over the past few millennia, about climate change, and the movement of glaciers, and about asteroid impacts on the Earth over millions of years, far longer than we have human records for and in fact even longer than humans have existed. The radioactive infant carbon that was created by cosmic rays impacting the Earth resides in every cell of every living being. How is all that connected?

Cosmic rays that propagate through air can cause electrons to be knocked from their orbits around atomic nuclei. In fact, many of the cosmic rays have enough energy to be involved in nuclear collisions; that is, the cosmic ray not only collides with the peripheral electrons, but also with the nucleus, located deep within the atmospheric atom's interior. This is no mean feat: whenever a cosmic ray gets close to the nucleus of an atom in the Earth's atmosphere, the positive electric charge that most of the cosmic ray particles carry interacts with that of the nucleus in the atmosphere, causing the two to repel each other, generally averting a nuclear collision. The particle and nucleus need to race toward each other head on at high speed to be able to overcome the repulsive force and collide. Nuclear scientists generally quantify the velocity needed for such collisions in terms of the energy of motion. That energy is often expressed in a unit that is useful when measuring the properties of individual protons, electrons, or other particles from the atomic world: the electron-volt, or eV. Such units have little meaning for us in everyday life because we deal with extremely large numbers of atoms in anything that we do. So in addition to using the unit that is commonly used by physicists, we will look at the velocities of the particles.

As an example, we follow one form of a cosmic ray particle, namely a fast-moving hydrogen nucleus, which is nothing but a solitary proton. In order for that proton to be able to collide with, say, the nucleus of an oxygen atom in the Earth's atmosphere, the proton needs to move at a speed of approximately 260,000 kilometers per second, or equivalently has to have an energy of motion of about 1 million eV. That required velocity is 87% of the maximum speed in the universe, the speed of light. Although such a speed seems pretty extreme already, it turns out that most cosmic-ray particles that reach the uppermost layers of our atmosphere move even faster than that. Because physicists need to express the punch packed by these particles, they prefer to use energies rather than velocities: as a particle's speed approaches the speed of light, energy and velocity no longer scale proportionally with each other as they do for the velocities we are used to dealing with. Very close to the speed of light, the velocity may go up only a little even as the energy of the particle goes up a lot. Thus, particles can have hundreds or thousands of times the energy needed to enable a nuclear collision, while still remaining under the universal speed limit set by the speed of light.

Cosmic rays reaching the Earth's atmosphere started outside the Earth's sphere of influence with much higher energies than needed for nuclear collisions. They have to have such high energies because the Earth's magnetic field very efficiently shields the atmosphere from all but the most energetic particles: as a cosmic-ray particle moves into the Earth's magnetic field, it is slowed ever more as it gets closer to the

Earth, where the magnetic field is stronger. Cosmic rays are electrically charged and have a hard time moving in directions perpendicular to a magnetic field, but can move much more easily along it. Thus, the efficacy of cosmic-ray shielding depends on the shape and strength of the planetary magnetic field through which the rays travel, and, therefore, on where on Earth and in which direction the cosmic ray is coming into the Earths' magnetic field.

The cosmic-ray shielding by the terrestrial magnetic field is least efficient right at the north and south magnetic poles, where incoming cosmic rays move mostly along the field. There, almost any cosmic-ray particle can reach the Earth's atmosphere by following the magnetic field that leads directly to the north and south magnetic poles. These poles are close to, but different from, the geographic poles, because the Earth's magnetic field is not quite aligned with Earth's rotation axis. When we look progressively further away from the magnetic poles, the energy needed to penetrate the Earth's magnetic field increases rapidly because particles have an increasingly hard time to approach the Earth as the angle of penetration relative to the field increases. For example, at latitudes of 65 degrees north (running through Alaska, northern Canada, Iceland, northernmost Scandinavia, and high-latitude Siberia) or 65 degrees south (far south of the southern tips of Chile and New Zealand), it already takes 100 times more energy to penetrate the Earth's magnetic shield and reach the atmosphere than would be needed for a nuclear collision. Near the equator, it takes about 140,000 times more energy to penetrate the magnetic field than would be needed for a nuclear collision. Even at those very high energies that are required to reach the Earth's atmosphere, there are still cosmic-ray particles that have that energy. Amazingly, some even have vastly more.

The most surprising property of cosmic rays is their diversity in energy. In outer space, cosmic rays range in energy from less than the energy needed for a nuclear collision to over a trillion times that energy. The most energetic particles are rare, however. They are, in fact, so scarce that there are only a few per year over the entire Earth. Overall, however, quite a few cosmic-ray particles are energetic enough to enter the Earth's atmosphere: near the equator, there is roughly one per square centimeter (more or less the size of a fingernail) every 10 seconds, while at magnetic latitudes above about 55 degrees (north of Denmark, the U.S.–Canadian border, and the Russian Kamchatka peninsula) there are a few hundred times more.

All of these penetrating cosmic rays induce nuclear reactions in the Earth's atmosphere. When such extremely energetic particles collide with atoms in the Earth's atmosphere, they simply shatter them into pieces. These atomic fragments include particles as exotic as their names: pions, mesons, neutrons, neutrinos, kaons, gamma rays, and so on. The energy of the initial cosmic-ray particle is divided over all of these fragments. Because the energy of cosmic ray particles often far exceeds the energy required to make the first nuclear collision possible, many of these fragments themselves carry enough energy away from the collision that created them to collide with other nuclei, in the process making more fragments, possibly creating cascades, or so-called showers, of nuclear fragments. The most common components of these showers that make it to near the Earth's surface are electrically uncharged neutrons. These are often captured when they collide with atomic nuclei because they are not bothered by any electric repulsion. They subsequently cause these atoms to change from one element to another as they eject electrons or protons or even small clusters of protons and neutrons while adjusting toward a more stable configuration. This process goes on until the cascade comes to an end, when the residual energies of all of the particles involved finally drop below the minimum needed for nuclear collisions.

Such particle showers were discovered in the late 1930s, but it took another 20 years before the detectors were powerful enough to determine how these showers worked in detail. Part of the investment in the development of such detectors was made in order to monitor manmade nuclear explosions, which first happened in the final phases of World War II. Networks of particle-measuring instruments were set up around the globe to detect both manmade nuclear explosions and naturally occurring cosmic rays and their particle showers. Now, a century after their discovery, we know that cosmic rays crashing into the Earth's atmosphere cause carbon-14 to be continually generated in nuclear reactions that occur in the very top layers of the Earth's atmosphere, called the stratosphere, particularly 6–10 miles (or 10–16 kilometers) above sea level.

Carbon-14 is not the only radioactive product formed in cosmic-ray showers. Others are beryllium-10 (^{10}Be) and chlorine-36 (^{36}Cl). These products are abundant enough, and survive long enough, for them to be measurable by modern-day physical techniques even in relatively old materials, including ice sheets and underground water reservoirs, or in the outer layers of rocks. Within a year or two of their formation, beryllium-10 and chlorine-36 find their way through the stratosphere to the Earth's surface, mostly because they become attached to particles of dust that themselves become trapped in precipitation in the form of either rain or snow. As such, beryllium-10 and chlorine-36 enter ponds, lakes, and rivers. These waters, in turn, percolate down deep into the Earth, into underground reservoirs or, in the polar regions, into lakes underneath glaciers. The presence of these elements provides a means to estimate, for example, how long water has been in an underground reservoir, by measuring the residual concentration of the radioactive compounds that continually decay away as they reside trapped within these deep basins shielded from cosmic rays.

Whereas beryllium-10 is not used in the human chemical network, chlorine-36 is: chlorine-36 mixes with the two stable isotopes of chlorine in the oceans. Chlorine combined with sodium, together extracted from the oceans, make the ordinary cooking salt that we consume. Some of the salt in our bodies was thus involved in cosmic-ray showers, just like the carbon-14 atoms. Very approximately, there are 500 billion chlorine-36 atoms in our bodies.

Anything on Earth is subject to these cosmic-ray particle showers, not only the high atmosphere. Some of the cosmic-ray particles are energetic enough to penetrate the entire atmosphere and to lead to nuclear reactions within the topmost layers of rock or soil. When the resulting products are radioactive, they are of use for geologists who want to determine how long rocks have been exposed to cosmic rays.

Suppose, for example, that there is a large meteor impact, or a large rockslide, or that faults or rifts are exposed during earthquakes. All of these processes suddenly expose rock surfaces, which are then subject to the effects of cosmic-ray showers that had not been able to reach them for many millions of years. Any of the radioactive isotopes with lifetimes of hundreds of thousands of years to a million years, say, will have decayed away while these layers of rocks lay hidden within the Earth. But once rocks are exposed to the cosmic rays at the surface, cosmic-ray showers begin to build up the concentrations of radioactive elements within the topmost 1 to 2 meters (3 to 6 feet).

Because we know the rates at which cosmic rays of a given energy penetrate the Earth's magnetic shield, and the efficiency of all nuclear reactions that occur within that atmosphere and within rocks of a given composition, as well as the decay rates of the radioactive components created by the nuclear reactions, we can deduce how long rocks and soils have been exposed to the cosmic rays. This was done, for example, for the large meteor crater in Arizona, which yielded an estimated age of some 50,000 years. This same radionuclide dating method is used to determine properties of glaciers during, for example, the ice ages: massive glaciers break off rocks from mountainsides and carry them toward their endpoints, where they accumulate in moraines. The time when that happened can be fairly accurately determined by measuring the difference in the radionuclide concentrations on their exposed topsides compared with those in the sides that face the ground and so have not been exposed to cosmic-ray showers.

The production rates of beryllium-10 and chlorine-36 in the stratosphere are very low: roughly one atom of beryllium-10 is created in the atmosphere per square centimeter per minute, and only one of chlorine-36 every nine minutes. In comparison, the production rate of carbon-14 is much larger: two atoms per second for each square centimeter. Overall, the balance between creation and decay puts about 100,000 tons of radioactive carbon-14 on Earth. Once carbon-14 atoms are formed either by the collisions of neutrons from a cosmic-ray shower with atmospheric nitrogen nuclei, or by the collision of cosmic-ray protons (hydrogen nuclei) with atmospheric oxygen nuclei, they behave just like any other carbon atom in the air: they generally bind with oxygen to form carbon dioxide. These carbon dioxide molecules circulate in the winds or move dissolved in the ocean currents around the Earth for some time before a tree, plant, moss, lichen, alga, or plankton organism absorbs it into its system. From there, it finds its way into the bodies of animals and people as they consume these other organisms as food.

Over the years, physicists realized that the production of carbon-14 continues in a very nearly steady manner from year to year. The steadiness of this production rate opens up a very useful way to use the decay of carbon-14 to date things: we can use the ratio of carbon-12 and carbon-14 to determine when living material turned into dead material, provided that we can measure this ratio accurately enough.

How does this carbon dating work? As long as plants, trees, or animals are alive, the cycle of carbon through their bodies continues. The carbon within them is continually exchanged with the carbon that exists in the Earth's atmosphere in the form of carbon dioxide. That supply of carbon dioxide carries a nearly constant fraction of carbon-14 mixed in with the much bigger bulk of carbon-12. When plant tissue eventually dies—for example, when a tree's living layers underneath the bark become core wood—no more carbon is exchanged with its surroundings. The same, of course, is true of animal or human tissue. Once a cell dies, the carbon-12 atoms in the organic molecules that are locked within it remain unchanged, but the carbon-14 atoms among them decay one after the other, turning into nitrogen. Thus, less and less carbon-14 will be left relative to the number of carbon-12 atoms. It is that changing ratio that allows us to measure how long it has been since something died and stopped exchanging its carbon atoms with its environment. The carbon-14 dating method can be used reliably for ages from a few decades to well over 20,000 years.

Carbon-14 decays not only when an organism is dead, but also during its lifetime. Decay is completely independent of where an atom is at the time: it can be in the atmosphere in carbon dioxide, it can be part of a sugar or protein or wall of a cell, it can be in an organic molecule inside an animal, or it can be anywhere inside the human body. Of all the carbon-14 atoms made by cosmic rays that are in our bodies, close to 4000 radioactively transition into nitrogen-14 every second. That adds up to a lot of transitions over a lifetime. Fortunately, our bodies manage to deal with our own natural radioactivity surprisingly well, with minimal detrimental effects.

Key Points: Chapter 8

- Every year, some 40,000 metric tons of material is added to the Earth by comets and interplanetary dust. That ancient stardust finds its way into our bodies, mixed with other matter that has been on the Earth for billions of years.

- Most carbon on the Earth endures for billions of years, but unstable carbon-14 atoms survive only thousands of years. Carbon-14 is radioactive and, over time, transitions into nitrogen. This is an example of radioactive decay.

- The unstable, radioactive state of carbon-14 proves that these are infant atoms, very much younger than all the other carbon on the Earth.

- Carbon-14, present in every cell of our bodies, links us to powerful cosmic rays from deep within the Galaxy. Cosmic rays that penetrate the atmosphere induce high-energy nuclear reactions when they shatter atoms in their path. This creates carbon-14 and other products, including chlorine-36, which enters our bodies when we consume salt.

• Living organisms use young carbon-14 atoms mixed with old, stable carbon-12 and 13. Carbon-14 decays at a steady pace and therefore the 14C/12C ratio can be used to determine the age of objects made of once-living tissue.

What's the use of a house if you haven't got a tolerable planet
to put it on?

HENRY DAVID THOREAU (1817–62),
in *Familiar Letters* (1865)

A man said to the universe:
"Sir, I exist!"
"However, replied the universe,
"The fact has not created in me
A sense of obligation."

STEPHEN CRANE (1871–1900),
in *War is kind and other poems* (1899)

The Origin of Elements

The Earth is close to 4.6 billion years old. The Sun must have existed for at least that long, because the large central mass of the Sun holds the rest of the solar system together, bound in its gravitational pull. Moreover, the Sun's brightness must have been relatively constant over the eons because otherwise the Earth would have been either too hot or too cold for the development or survival of life as it now exists. All of these life forms are present either in liquid water or in environments with enough of that all-important fluid nearby. Much of the Earth, therefore, must have been warm enough to avoid glaciation and yet cool enough to avoid desertification for a very, very long time.

Some 3.9 billion years ago, there appears to have been an era in which very large meteor impacts occurred frequently, possibly all over the solar system. Life before that period may have been possible, but it would mostly likely not have survived the impacts. Starting from the end of what is referred to as the "late heavy bombardment", computer calculations based on the laws of stellar physics show that the Sun's brightness has been steadily increasing over time, but by no more than about a third over the 3.9 billion years of its existence as a stable, mature star with a fully formed Earth orbiting around it. Although this may seem substantial, from an astrophysical perspective this increase in brightness is only slight, especially given that the Sun will light up by more than 1000 times its original brightness at the end of its lifetime. This will not be happening soon, however, as the Sun has another 4.5 billion years to go before it even enters that phase of major brightening.

The limited increase in the Sun's brightness over billions of years means that the Earth has stayed within what is known as the habitable zone. That zone represents the range of distances from the Sun within which life as we know it is possible, at least in principle. One definition of the habitable zone expresses it as the distance range from the Sun in which an Earth-like planet can have liquid water on its surface. Closest to the Sun, that zone ends where the sunlight is so intense that the energy-trapping effects of the planetary atmosphere would be too strong: the resulting hot atmosphere would cause the water vapor contained in it to simply evaporate off the planet. Depending on its chemical composition, the atmosphere of a planet too close to a bright star like the Sun could become so hot that even rocks on the surface would melt. For the present-day Sun, and with our present-day understanding of planetary climates, the inner edge of the habitable zone is thought to be located only a little inside the Earth's orbit around the Sun.

The outer edge of the liquid-water habitable zone lies where the planetary surface temperature would be so low that surface water would freeze. It turns out that the distance from the Sun of that outer edge is not easy to estimate because it requires a thorough understanding of all the processes that contribute to atmospheric warming under the Sun's glow. These include heat trapping within the atmosphere on climate time scales of decades and more, as well as the chemical evolution of the atmosphere as it interacts with the planetary crust on geological time scales. Current estimates of the outer edge of the habitable zone range from about 1.4 times the Sun–Earth distance to maybe 2.4 times.

A much narrower range of a habitable zone emerges when we ask the question differently. Instead of asking where liquid water can exist, we could ask where enough carbon dioxide would exist in the atmosphere of an Earth-like planet. After all, plants, on which most life on the Earth's surface and in the surface layers of the oceans relies, can thrive only with a minimum level of carbon dioxide in the atmosphere. That minimum plant-sustaining concentration is estimated to be about 10 parts per billion (compare that with the current level in the Earth's atmosphere of about 390 parts per billion).

Carbon dioxide is removed from the atmosphere by chemical weathering, when carbon dioxide dissolved in rain comes into contact with rocks. The rate of such weathering goes up with increasing temperatures, and so climate and atmospheric composition are coupled. Volcanoes intermittently replenish the carbon dioxide that is gradually weathered away by injecting carbon dioxide from the subsurface reservoir into the atmosphere. Should the trapping by weathering increase because of increasing temperatures at a given level of volcanic action, however, then the net amount of carbon dioxide in the atmosphere would decrease, eventually causing the temperatures to go down again. The opposite will happen if the temperature decreases. This coupling feedback considerably narrows the range of distances from the Sun within which a planet can support photosynthesis. By some estimates, the continued slow brightening of the Sun will eliminate the possibility of photosynthesis for our planet some 1 to 2 billion years from now.

Because the habitable zone in the solar system is not particularly wide, the unavoidable conclusion is that the Sun cannot have changed its brightness by much over the time that life has existed on the Earth: otherwise, the narrow habitable zone would have easily shifted away from the Earth's orbit, leaving the Earth a dead, barren planet. Yet we know that bacterial life on Earth goes back at least 2.5 billion years (which is the age of the oldest microfossils), likely as much as 3.5 billion years (which is the age of the oldest microfossil-like finds that suggest the existence of cyanobacteria), and perhaps even 3.9 billion years. With that information, we would infer even without knowledge of the physics of stars that the Sun has not changed its brightness by more than a few dozen percent over the past 3 billion years.

Yet, without knowledge of stellar physics, we would not understand how our comfortably warmed home planet in cold, inhospitable space could have been illuminated for that long by a fairly constant Sun. Before the beginning of the twentieth century, the magnitude of that problem was not realized, because the Earth's age was estimated to be considerably less than we now know it to be. Even then, however, researchers struggled to understand how the Sun could sustain its brightness for as long as they thought. One thing they established fairly early on is that the energy radiated by the Sun could not come from chemical reactions. For example, even if the entire Sun were to consist of carbon and oxygen in the correct ratio, the burning of that carbon (as with coal, for example) to fuel the shining Sun could not maintain the Sun's glow for more than approximately 50,000 years. They then thought of another energy reservoir that is far larger than what can be supplied by chemical reactions. That energy would be associated with the Sun's gravity. If all of the Sun's gas could collapse into the center of the Sun, enough energy would be released to power the Sun's brightness for as long as 30 million years. That is 600 times longer than what chemical reactions could sustain. Compression of all that gas into a very small volume in the center of the Sun is hard to imagine, but in any case even that estimate falls short of what is actually needed to power the Sun over the true lifetime of the Earth by a factor of at least 150.

For a long time, astronomers struggled to find the real source of energy that powers the Sun. In groundbreaking work spanning multiple decades around the end of the nineteenth century and the beginning of the twentieth century, the pieces were finally assembled: the discovery of radioactivity, the development of the theory of relativity, and the realization of quantum mechanics had to come together before the evidence unequivocally pointed to nuclear fusion. With these three pieces of knowledge solidly in place, nuclear fusion was demonstrated to be the primary source of stellar energy. Getting all the numbers of the theories to match the observations was by no means easy and did not happen until after several decades of interactions between nuclear physicists and astrophysicists, but the knowledge that emerged from scientific debates resulted in a thorough understanding of stellar interiors. As it turned out, a star like the Sun has enough hydrogen to fuse into helium to shine as it now does, for a total of about 10 billion years.

Where does the story of the Sun and the planets begin? We now know that stars form out of the vast clouds of gas that float around in any galaxy. The atoms, molecules, and dust particles in such clouds sense each other's gravity and, as a result, begin to move toward each other. Over time, the atoms, molecules, and dust particles move toward each other ever faster in an ever-denser cloud. They cannot avoid colliding with other particles trapped in the same process of contraction. In these collisions, they bounce around in random directions even though there is a net motion toward the center of the cloud. As the atoms of a gas move about with increasing speed, the temperature of the gas (which is an alternative expression for the average speed) rises. However, when the density and the temperature of a gas increase, all the bouncing atoms and molecules in it exert mounting pressure on anything in the neighborhood. That pressure slows and, once sufficiently large, halts the contraction of the gas cloud.

As long as the cloud is in a phase where it is not very dense, the gas in it can cool down fairly efficiently by radiating energy away: the denser parts glow in infrared light, and the energy lost in that limits the increase in temperature. When the cloud has contracted to the point where it has become very dense, however, this glow can no longer easily escape into galactic empty space. Then, the light runs into the particles of the cloud before it reaches its edge, is reabsorbed, and is therefore essentially stuck within the cloud. Once the gas density exceeds a critical value, the contracting cloud can no longer counter the increase in temperature that was a direct consequence of the contraction. Then, both the pressure and the temperature in the deepest interior of the collapsing cloud rise rapidly. Eventually, the pressure becomes so large that it slows the contraction down to an ultimately very gradual process that lasts millions to tens of millions of years, depending on the initial mass of the cloud.

Even though the heated interior of a contracting volume of galactic gas can slow the gravitational collapse, it cannot stop it until nuclear fusion occurs. Only that fusion can keep the temperature and pressure high by supplying new energy in the core to match what is lost from the glowing surface. Fusion comes about because the collisions between the gas particles deep inside the cloud ultimately become extremely violent, allowing atomic nuclei to merge in the process. Collisions between the nuclei of hydrogen atoms—protons—are the first result of this. Proton–proton nuclear collisions result in something that is unstable. The two-proton body that temporarily forms either immediately falls apart again, or a positron is ejected from it, which causes one of the protons to transform into a neutron. The latter process results in what we know as "heavy hydrogen", or deuterium. With the gas cloud still contracting, the deuterium nuclei collide with other hydrogen nuclei, resulting in a light form of helium that has nuclei containing two protons and one neutron, which is known as helium-3, or ^{3}He. Eventually, the light helium nuclei successfully merge with each other to form the most common form of helium, with two protons and two neutrons in its core. In the end, four hydrogen atoms are thus fused into one helium atom.

In this fusion process, which occurs in the Sun as in every other star, another type of fundamental particle is involved: the neutrino. Neutrinos are very peculiar, in that they rarely interact with other matter. The interactions are so infrequent, in fact, that almost all of the neutrinos that are created in nuclear fusion inside the Sun today easily escape from it without difficulty, despite all the overlying layers of gas between the solar core and interstellar space. For the Sun, the loss of energy by neutrinos has little impact, but in the terminal phases of stars much heavier than the Sun they turn out to be very important because they are the particles that lead to the explosion of these stars. By their violent deaths, stars lose their gravitational grip and release the elements from which we are made, as we discuss later in this chapter.

Once the nuclear fusion in the contracting gas cloud gets going, the reaction rate quickly increases until it provides as much energy as is radiated away from the surface. At that point, continuing collapse is avoided and a star is born: a huge cloud of galactic gas has now transformed into a dense, hot nuclear furnace in which hydrogen is fused into helium just fast enough to ward off a catastrophic collapse for millions of years. Modern-day computations of stellar interiors tell us exactly how long hydrogen fusion can counter gravity. The mass of the star is the primary factor that determines the duration of this phase. Heavier stars need a higher pressure to counter their own gravity, which means higher collision rates in their interiors, and therefore a higher nuclear reaction rate, resulting in a brighter star. The brightening is disproportionate to the mass, though: heavier stars run through their hydrogen fuel faster, even though they have more of that resource available. For example, a star like the Sun will be in its prime, smoothly fusing hydrogen into helium, for over 9 billion years. A star twice as heavy, in contrast, burns through its hydrogen reservoir 10 times as fast, while a star 15 times heavier goes through it more than 800 times faster. In other words, heavy stars shine brightly, but in order to do so, they devour their hydrogen far, far faster than our Sun. On the other side of the mass scale are the lightweights among the stars, which use hydrogen more sparingly and therefore last for a longer time. For example, a star with a mass only 20% less than the Sun can sustain hydrogen fusion for an estimated 20 billion years, which is more than twice as long as the Sun can.

The age of the universe is presently estimated to be about 14 billion years. That means that any star heavier than approximately 90% of the weight of the Sun cannot have been fusing hydrogen since the beginning of time. Heavy stars that formed very early in the life of the universe no longer exist today. Even the Sun itself, though it is now only halfway through its hydrogen reserves at 4.6 billion years of age, must have formed at a time when the universe as a whole was already 10 billion years old. Stars that now exist that are 15 times heavier than the Sun must have started fusing hydrogen even more recently, because these heavier stars can last only for less than a tenth of the current age of the universe. Stars heavier than that have even shorter life spans. Consequently, none of the relatively heavy stars that we now see in the sky—particularly the bright blue heavyweights—can have been around for long, and did not even exist yet in the relatively recent age of the dinosaurs, which ended 65 million years ago. By the laws of nuclear astrophysics, no currently visible star heavier than roughly four solar masses can have been shining in the Earth's night sky when the American and African continents drifted apart. The oldest possible eyes on Earth, which might have existed when multicellular life came about some 600 million years ago, might have witnessed only those stars among the ones we now see that were less than three times as heavy as the Sun.

Despite the fact that stars as heavy as the Sun or heavier cannot have survived in the universe from the beginning of time, we see stars that are much heavier than the Sun in the sky. Surely, this cannot mean that heavier and heavier stars formed later and later in the life of the universe just so that humanity could see them shine all at the same moment in time. Instead, it means that stars are made over and over, so that at all times there is a mix of very heavy, heavy, intermediate, and light stars around. In order to be able to keep making stars for such a long, long time, somehow new volumes of gas must be around out of which new generations of stars can be made. Nature's way of doing just that is to blow up the heavier stars at the end of their lives, enabling the cycle of star formation to start over. It is not quite a perfect cycle, however. The material that is thrown back into the Galaxy when a star explodes is enriched with heavier elements as a result of nuclear fusion. Much later, out of that material, solar systems with planets such as ours can be formed.

After a star has converted most of the hydrogen in its core into helium, it can continue to maintain itself against the pull of gravity for a bit longer by fusing the hydrogen that lies higher up, above the deep interior region that is, by then, all helium. Fusion continues to move outwards where hydrogen fuel is still available. When a star reaches this phase, things accelerate. Even as the hydrogen-fusing shell propagates outward, the core contracts under its gravity until it is dense enough for helium to fuse into heavier elements, including carbon, nitrogen, and oxygen. For a star like the Sun, that is as far as fusion can go, but a substantially heavier star can continue this process, fueling itself by working through increasingly heavier elements, all fusing in outward-moving layers, one after the other. Eventually, a heavy star resorts to creating elements such as neon, magnesium, silicon, and sulfur, until iron is created. Every one of these steps proceeds faster than the previous one, mostly because each of these fusion steps liberates less energy than the last. For the star this is problematic because it becomes brighter over time as it expands its outer layers and has more surface area to lose energy from. In a star that is 25 times heavier than the Sun, for example, hydrogen fusion lasts for some 7 million years, carbon fusion for 500 years, oxygen fusion for a mere five months, and silicon fusion into iron terminates within a day.

Iron has the most tightly bound nucleus of all the elements, so that fusing elements heavier than iron requires more energy than is brought together by the colliding nuclei. When most of the star's core consists of iron (with a smattering of heavier elements created as side products), surrounded by layers of increasingly lighter elements, no further viable nuclear reaction pathway exists for substantial thermonuclear fusion. At that point, the gas pressure loses its battle against gravity and the star simply caves in on itself.

For a moment, however, let us step back from the catastrophic collapse that threatens the star's existence, because several important bits of knowledge were skipped until now. First of all, the majority of the stars in the sky have less than 90% of the Sun's mass. These lightweights are energy efficient, and the lightest among them can sustain hydrogen fusion for tens of billions of years, much longer than the current age of our universe, which is an estimated 14 billion years. Thus, none of these stars could be old enough to have run out of hydrogen, because the universe has not existed long enough for them to use it up. Yet, there are elements heavier than helium in their interiors, which we recognize by looking at the colors of their light, using spectrographs, which unravel light in any desirable detail. Where did those stars get those heavier elements? While we are dealing with that question, we also have to wonder why some stars have these heavier elements in abundance, while others have hardly any at all.

A crucial clue to helping us answer the question regarding the abundance of certain elements in stars—which will also tell us why those elements exist in our bodies—lies in tracing the elements as a function of the ages of stars. Here, it helps us that not all stars are loners. In fact, most stars exist as pairs, while others come in clusters of dozens, hundreds, or even millions. When we observe compact clusters of stars with all their member stars close together, we can assume that those stars were formed at about the same time and have approximately the same age. As such a cluster ages, the heaviest, brightest stars run out of fuel first and disappear from sight. Over time, there is always a heaviest surviving star that stands out as the brightest of its peers, before it, too, has to gasp for the final available fuel. The older the cluster, the fainter and cooler will be such a brightest senior star. Astronomers can rank clusters in order of their age by observing the brightness of the stars as well as their colors, which are directly related to their temperature: a hot star is blue, a warm star like the Sun is white or yellow, and a cooler star is orange or red. With the help of the computations of nuclear physics, astrophysicists can assign an actual age to the clusters with pretty good accuracy.

Comparing the age of cluster stars with the abundance of their elements, such as carbon, nitrogen, oxygen, and iron, reveals that there is a very clear trend for younger stars to have more of the heavy elements. The oldest stars are more than 13 billion years old, which makes their age close to that of the universe. These oldest stars are made up almost entirely of hydrogen and helium, and the heavier elements are thousands of times less abundant than in the Sun. What does that mean? Somehow, the universe around us must have started out consisting of almost exclusively hydrogen and helium, while anything heavier—and thus much of what our bodies are made of, and most of what our planet is made of—came into existence later.

Our current understanding of the universe is that it began with a "big bang": an explosion from an initial state that was incredibly dense and exceedingly hot. There is a diversity of convincing evidence in support of the Big Bang theory. One signature of the Big Bang is its faint afterglow, which used to be hot, but which has by now cooled to a temperature very close to absolute zero. That very weak infrared glow has been detected from all directions using extremely sensitive instruments. Another piece of evidence that supports the Big Bang theory is the fact that galaxies all around us continue to move away from us, continuing the expansion of the universe that was started by its explosion. Galaxies are moving away from each other in the expanding universe much like raisins in rising dough or like the markings on the surface of an inflating balloon.

Starting as it did being very hot and dense, the universe went through a phase in which it, itself, was a nuclear furnace in its entirety. Cosmologists and nuclear physicists argue that the phase of nuclear fusion in the early universe lasted a mere 17 minutes, from 3 to about 20 minutes after the beginning of time. In those first minutes, hydrogen formed first and approximately one-quarter of it was fused into helium. In the process, about 2 hydrogen nuclei in 1000 merged into deuterium (the heavy form of hydrogen, containing one proton and one neutron), 3 hydrogen nuclei in 10,000 formed a light form of helium (missing one neutron relative to common helium), and less than 1 in 100 million formed lithium. Of the other elements, negligible amounts were formed.

After this 17-minute phase of nuclear fusion of the very first elements, it took almost half a million years for the universe to cool down sufficiently for negatively charged electrons to bind to the positively charged nuclei to form the neutral atoms that we are familiar with on Earth. It probably took another 100 million years for the first stars to form in the new universe, which was filled with gas clouds all over the place. That first generation of stars would thus have been formed from that original matter, with no measurable carbon, nitrogen, oxygen, or other elements. It is in the cores of those stars, over millions to billions of years of nuclear reactions, and in the ultimate explosions of the most massive among them, that all the heavier elements are fused into existence.

Newly formed heavier elements clearly did not stay locked up inside the earliest stars, but must have become available for the formation of later generations of stars. Otherwise, the heavier elements would not exist in all the stars around us. How did the fused material find its way into the multitude of galaxies that we now observe to be filled with stars with heavier elements? Decades of research—observational, experimental, and computational—have revealed much of that story. Many of the details require further research, but we do now have a rather comprehensive picture of what happened after the first-generation stars ended their mature phase.

As a star runs out of hydrogen in its core, the core loses a little of its battle with gravity, and contracts. As we described earlier, nuclear fusion then moves to a shell around the core, so that hydrogen fusion can continue for a while as that shell moves outward, leaving almost pure helium in its wake. At some time, the core becomes dense enough for helium to begin to fuse, which, through a series of nuclear collisions, results in the creation of carbon. This uses up helium underneath the hydrogen-fusing shell faster than that shell is moving outward. Therefore, at some point in time, the helium fusion runs close behind the hydrogen-fusing shell. The heat from the core pushes that shell upward, making it less dense and thereby slowing its rate of hydrogen fusion. That is not good for the star's longevity: the star continues to shine its energy away, while less than what is needed to match that loss is being generated in the interior.

Next, an interesting and important interplay happens between different interior domains of the star. The helium-fusing core expands into an area of the star where there is less helium, because hydrogen fusion has not yet run its full course there. When helium fusion slows down as a result of scarcity, the deep interior contracts and loses some ground against the pull of gravity. This contraction, however, reignites the hydrogen-fusing shell around it, creating new helium there that can then be fused in its deepest, densest, hottest layers. This pushes the hydrogen shell outward, slowing its fusion once again. Depending on the mass of the star, this stop–expand–contract–restart cycle can be repeated dozens of times and over many millennia. In each cycle, the star swells up and then shrinks again, like lungs gasping for air, albeit on a vastly grander scale. In the later phases of this process, the expanding motions become so fast that they actually catapult material off the star into distant interstellar space. Throughout this period in the life of the star, the motions of gases in its interior cause gas from deeper layers to mix into higher ones. Thus, higher layers are enriched with the heavy-element fusion products from deeper down, including carbon, oxygen, nitrogen and, in heavier stars, also neon, sodium, magnesium, and fluorine. The outer layers that are blown off in a stellar wind, if not thrown off the pulsing star, thus eject heavier elements into the interstellar gas that surrounds the dying star.

In the final throes of this phase in a star's life—lasting less than 1% of its overall lifetime—much of the outer part of the star is eventually lost into interstellar space. Estimates suggest that as much as one-third of all of the carbon in our Galaxy, for example, was thrown off stars during this phase. The nuclear reactions during this pulsating phase generate lots of free-flying neutrons, which, one by one, merge into atomic nuclei, making elements even heavier than iron. Perhaps half of the total mass of elements heavier than iron in our Galaxy is created in, and subsequently expelled from, stars nearing their demise.

All stars heavier than about 90% of the solar mass reach their final phase of their existence within at most 14 billion years. The heavier the star, the faster it reaches it. Hence, multiple generations of the heavier stars have contributed to the buildup of elements beyond hydrogen and helium in the galactic interstellar gas. All stars, up to those eight to nine times heavier than the Sun, end their lives this way. With much of their mass lost through the pulsations, and with little fuel left to keep them warm enough to battle gravity, these stars begin a long period of contraction, eventually changing into "white dwarfs". They are called "white" because for a long time they literally glow white hot. And they are called "dwarfs" because they become as small as the Earth (100 or more times smaller than their original size as adult stars). Then, quantum-mechanical forces that we are unfamiliar with on the Earth halt the gravitational contraction of these stellar cinders. This is, in fact, also a preview of the future of our own star: some 4.5 billion years from now, the Sun will go through such a phase of expansion and pulsation, destroying Mercury, Venus, and likely the Earth in this phase. Then it will become a white dwarf, glowing ever fainter and redder as it cools.

Stars heavier than approximately eight solar masses have a final surprise in store. When these stars run out of fuel for fusion in their cores—that is, when their cores finish fusing silicon into iron—nothing can stop the gravitational collapse, until matter becomes as dense as the nuclei of atoms themselves. That cataclysmic core collapse happens in a matter of seconds. It can happen so quickly, in part, because energy escapes from the stellar core in the form of neutrinos: as the matter becomes denser, all sorts of nuclear reactions happen that utilize energy as they form elements heavier than iron, with some of that energy radiating out in the form of unstopped neutrinos. Once the matter becomes as dense as atomic nuclei, the collapse suddenly stops. That is the beginning of the spectacular end of the star in a magnificent explosion. What, in detail, forces the explosion remains to be understood. It may be a shock wave starting at some point during the collapse and eventually moving outward, or it may start with the corresponding changes in the overlying layers, or perhaps it is the effect of neutrinos leaking energy as they move unimpeded out of the interior, or some combination of these possibilities. Whatever the detailed cause, in that explosion most of the star is blown apart in what we call a supernova.

A supernova is a remarkable spectacle, best witnessed from an enormously great distance. The dying heavy star throws out many solar masses of material, with a brightness that may exceed that of millions of stars together, sometimes briefly exceeding the brightness of all the stars of an entire galaxy combined. Within the explosion, nuclear reactions of all kinds are happening. Vast numbers of the newly fused atomic nuclei are unstable and decay radioactively, some in a fraction of a second, some in days to weeks in the supernova afterglow, and some in the subsequent billions of years. During the first weeks to months of the supernova, radioactive decay powers much of the expanding cloud's glow. The supernova destroys many of the previously created heavier elements but newly creates many others. A lot of the elements that are heavier than iron, including the ones that we know as naturally occurring radioactive elements on the Earth, are created in these truly astronomical explosions.

New stars and their planets, including ours, can be formed out of that cosmic debris. Some of the longest-lived radioactive elements in the exploded mix survive for billions of years. Among them are uranium-238 (with a half-life of 4.47 billion years), thorium-232 (with a half-life of 14.05 billion years), and potassium-40 (with a half-life of 1.25 billion years). The three longest-lived radioactive isotopes on the Earth, embers of a supernova at least 4.6 billion years ago, survive in substantial quantities even today: the decay of these elements (potassium-40, thorium-232, and uranium-238) continues throughout the Earth. In fact, the decay of potassium-40 is the dominant source of the natural low level of radioactivity that is always present in our bodies, with thousands of decays each second of our lives. These radioactive decays are slightly more frequent than those associated with carbon-14 in the body.

For the billions of years of the Earth's existence, the energy released by the fission of long-lived radioactive matter deep within the Earth has helped keep the internal temperature of the Earth higher than it would have been otherwise. The high temperature of the interior keeps much of it liquid, and that in turn powers the Earth's dynamo that generates its magnetic field. This field not only shields us from many of the dangerously fast cosmic-ray particles from outer space, but also helps in maintaining the Earth's atmosphere. Without the radioactive materials formed in multiple supernovae prior to the formation of the solar system, the Earth would not have a hot core, and therefore would have been left without a magnetic field, or at least would have a much weaker one, with catastrophic consequences. Without that field, our atmosphere would have been efficiently stripped from the Earth by the solar wind, as has happened in the case of Mars, thereby eliminating the possibility of all life as we know it. This is one of the many conflicting lessons from the universe: a nearby supernova may wipe out life on Earth, but without the supernovae that came before, we would not have existed.

By the way, the third most common element in the Earth's atmosphere by weight, after nitrogen and oxygen, is argon. Of all the argon in the Earth's atmosphere, 99.6% is the isotope argon-40, which differs distinctly from the most common argon isotope in the Galaxy. Outside the Earth's atmosphere, stable argon-36 is the most common form. That has four fewer neutrons in its nucleus than does argon-40. Argon in the wind blowing from the Sun, for example, is 85% argon-36. The bulk of the argon-40 in the Earth's atmosphere therefore cannot come from the gas from which the solar system was formed. Instead, we now realize, it is the result of the radioactive decay of one of those unstable isotopes formed in past supernovae, namely the decay of radioactive potassium-40 into stable argon-40. Thus, most of the argon in the Earth's atmosphere is a chemical that is much younger than the universe, and in fact younger than the planet itself. It is only because argon is a noble gas, chemically reacting with hardly anything, even in laboratory conditions, that it is not part of the story of the chemicals in our body. Nevertheless, even though our body does not process it, we breathe in the decay product of radioactive supernova embers all the time, in concentrations that, for argon alone, are 26 times larger by volume than carbon dioxide.

Another decay product of potassium-40 is, however, part of the story of our bodies: this is calcium-40, which also occurs in abundance elsewhere in the universe, formed in stars and their supernovae. Yet, part of the calcium that forms our bones was created after the formation of the planetary system, when supernova remnants buried within the Earth finally transitioned from an unstable to a stable nuclear form.

We now know that stars are the nuclear furnaces of our universe. They can exist because they thrive on nuclear fusion, which transforms the universe's original elements, hydrogen and helium, into all the heavier elements that exist in space, on the Earth, and in our bodies. After more than 14 billion years of aging, pulsating, and exploding stars, the interstellar material around a star like the Sun would contain all of the elements that occur naturally on the Earth, including the radioactive ones. As a consequence of all this nuclear and astrophysical activity around where the solar system originally formed, the ages of the elements in our bodies have a wide range. Hydrogen has been around for the full 14 billion years of the universe's existence. Almost all of the heavier elements were formed sometime between the age of the universe and the formation of the solar system, 4.6 billion years ago. Part of the calcium in our bones came into existence at some point during the existence of the Earth through radioactive decay. Only the isotope carbon-14 was made in recent history, mostly within the past 10,000 years or so, starting with particles escaping from supernova shocks far away in the Galaxy and ending in a nuclear reaction somewhere in the Earth's stratosphere.

Key Points: Chapter 9

- The source of energy for the Sun and all other stars is nuclear fusion. Heavy stars use much more energy and exist for a shorter time than light ones, but all stars begin and end. A middleweight star like the Sun can continue nuclear fusion for about 10 billion years.

- Stars form out of vast clouds of gas. Cloud particles fall toward each other subject to gravity, eventually igniting nuclear fusion, causing a star to be born.

- Nuclear fusions in stellar interiors form successively heavier elements. When a Sun-like star is nearly out of fusable material, it cycles through many expansions and contractions, during which heavy elements from its deep interior mix into the original outer layers, much of which are eventually blown off into interstellar space.

- Most stars expire by permanently caving in onto themselves, but those more than eight times heavier than the Sun terminate in spectacular supernova explosions.

- All elements in our bodies heavier than hydrogen were formed by nuclear fusion inside stars. They have been ejected from dying stars throughout the history of the universe. We are made of elements with an age range of 4.6 to 14 billion years, mixed with a sprinkling of much younger isotopes, such as carbon-14.

- **Stars are the nuclear power plants of the universe. They keep going on nuclear fusions, first of hydrogen, the universe's original element, while later on using heavier elements created in their interior.**

The universe is wider than our views of it.

HENRY DAVID THOREAU (1817–62),
in *Walden* (1845)

In order to fix in our minds the vastness of the system that we shall have to consider, I will give you a "celestial multiplication table". . . .

A hundred thousand million Stars make one Galaxy;
A hundred thousand million Galaxies make one Universe.

The figures may not be very trustworthy, but I think they give a correct impression.

ARTHUR STANLEY EDDINGTON (1882–1944),
in *The Expanding Universe;
Astronomy's "Great Debate"* (1933)

10

Cosmic Rays and Galactic Ecology

The cosmic rays that cause the young, radioactive carbon-14 and chlorine-36 to exist within all living tissue on the Earth link us directly to both the Sun and to our Galaxy, the Milky Way. The story of cosmic rays thus tells us part of the story of our bodies. In addition, it comprises the final chapters in the formation of all the elements, before they came to shape the Earth.

Scientists continue to complete the fascinating story of the origin of cosmic rays even now, over a century after their discovery. Within five years of the discovery of radioactivity by Henri Bequerel in 1896, it was known that the radiation given off by radioactive substances came in three basic forms: alpha, beta, and gamma. Ernest Rutherford, who coined these terms, had established that the alpha particles produced by naturally occurring radioactivity could be stopped by as little as a sheet of paper in front of the radioactive source. Beta radiation had a greater penetrating power, but even that did not travel through the thin matter of air by more than a few feet. In 1900, Paul Villard discovered a third type of radiation, which Rutherford suggested be named gamma radiation. Gamma radiation penetrated much further than alpha and beta radiation, and it was thus natural for these early radioactivity researchers to assume that cosmic rays, which can travel through several kilometers of air, would correspond to that most penetrating of radiation then known. That assumption, however, was wrong.

The nature of alpha, beta, and gamma radiation was investigated using all the tools available at the time. Among them was equipment designed to find out if the radiation was sensitive to electricity and magnetism. Whenever an electrically charged particle travels through an electrical or magnetic field, its trajectory is bent as it responds to the electromagnetic force. Alpha and beta radiation are deflected in opposite directions by electromagnetic forces, and hence were shown to have opposite electrical charges: positive for alpha, negative for beta. Gamma rays could not be deflected by either magnetism or electricity, so it had been concluded that whatever these rays were made of, it could not have an electrical charge. Studying radiation and its sources in a laboratory, however, is quite different from having to deal with the elusive cosmic rays coming from somewhere outside the reach of laboratory researchers.

The investigation of cosmic rays required other strategies, including the use of forces available in nature, associated with the entire Earth. When Jacob Clay measured cosmic rays during his sea voyages between Italy and Indonesia in 1927 and 1928, he noted that the intensity of cosmic radiation depended on the latitude at which he made the measurement: the closer he came to the equator, the lower the intensity. He was uncertain what to make of that finding. The position dependence of cosmic-ray intensities might somehow reflect differences in the Earth's atmosphere between the high latitudes of central Europe and those of the equatorial regions that he traveled through.

A careful study of the details of the variations discovered by Clay provided a clue that cosmic rays were, in fact, electrically charged and therefore could not be gamma rays. This was definitively proven in 1933, when cosmic rays were measured at many locations around the world. It became clear that it was not simply geographical latitude but rather the geomagnetic latitude—the distance to the Earth's magnetic poles— that determined the intensity of cosmic rays. The mismatch between the Earth's magnetic poles and its geographic poles changes slowly with time. In the year 2000, for example, the magnetic north pole was located just northward of 80 degrees geographic latitude in the northern reaches of Canada. Around 1933, it was even further off geographic north, lying somewhat west of Resolute, Canada, close to a geographic latitude of 73 degrees north. Thus, it was even easier then than now to see that the cosmic ray intensities depended on the magnetic field.

It took until 1941, almost 30 years after Victor Hess's balloon flights described in Chapter 8, before it was determined that cosmic rays contained mostly protons, which are the positively charged nuclei of hydrogen stripped of their single orbiting electron. Another three decades later, it was discovered that cosmic rays are not only made of isolated hydrogen nuclei flying around, but that they also contain a rich mix of all the elements that exist around us in the cosmos. The reason why these atoms of commonly occurring elements have such strong penetrating power is that they are moving incredibly fast, approaching the speed of light. No single collision with the atoms in the atmosphere can stop such a fast particle, even if that collision shatters the nuclei of the atoms involved. The colliding particles and their fragments share the large energy of the original cosmic-ray particle, which still leaves each of them with far more than can be stopped in a subsequent collision. Generally, each cosmic-ray particle entering the Earth's atmosphere involves a chain of collisions, causing a cascade of up to billions of particles, including protons, neutrons, electrons, and fragments of heavier elements.

Although the space between the planets and the stars is often referred to as "vacuum", it is not altogether empty. Hence, in order to reach the Earth's atmosphere, the cosmic rays must have traveled through that material first. The Earth's atmosphere is not very dense, but nevertheless there is enough to put a total of about two pounds (or one kilogram) of atmospheric mass above each square centimeter (or about a quarter of a square inch) at sea level. This begs the next question. If cosmic rays have difficulty penetrating the Earth's atmosphere, which is only a few kilometers (or miles) thick, then how far could they fly through interstellar space before having a similarly difficult time penetrating all that floats between where they originated and our Earth?

It turns out that interstellar space contains so little matter that cosmic rays can travel through much of the Galaxy, some 100,000 light years across, without suffering too much from the interactions with the gas, dust, and planets between the stars. Each light year is the distance that light can travel in a year, which is about 1000 times the span of the solar system, so cosmic rays can certainly go a long way before they collide with other particles in our Galaxy.

If the cosmic rays that enter our atmosphere can travel through much of an entire galaxy, could they come from outside the Galaxy, from other galaxies, or even somehow from the gas between the galaxies, the intergalactic matter? Although possible in principle, it appears that we do not have to look quite that far for the source of most of the cosmic rays that reach the Earth. The evidence for that comes from a variety of observations combined with computations, but the basics can be understood by looking at only two key pieces of that evidence.

The first key piece comes from the fact that cosmic rays are influenced not only by collisions with the few particles floating around in interstellar space but also by the magnetic field that exists everywhere within it. Being charged particles, they can travel freely along the magnetic field but can only spiral around it in a direction perpendicular to the field. A magnetic field exists throughout the Galaxy. The sources of that magnetism are variable and all impart their own preferred directions, so that the combined galactic field has all manner of directions, strengths, and patterns of change, depending on position. The result is that wherever a cosmic ray particle reaches a point where the magnetic field changes in direction or strength, the particle is deflected in another direction. As a result of the chaos in the collective magnetic fields, these particles execute a random walk around the Galaxy, rather than flying in a straight line through, and—unless colliding with something— out of the Galaxy. The estimated time that cosmic rays are effectively trapped within the Galaxy, wandering about randomly before they escape when, by chance, they reach the Galaxy's outer edge, is something close to an average of 15 million years. After that they arrive at the ends of the reach of the galactic magnetic field and leak away from the Galaxy, into the vast emptiness of intergalactic space.

The second piece of evidence needed to show that cosmic rays originate mostly from within our Galaxy is provided by a byproduct of the cosmic rays. While wandering about our Galaxy, as they do in every other galaxy where they are similarly created, cosmic rays do occasionally collide with interstellar atoms. As in the Earth's atmosphere, these are highly energetic collisions that can shatter atomic nuclei. When this happens, one of the byproducts can be a gamma ray, which is a very energetic form of light that can be captured by special cameras on spacecraft. The images made with these cameras show that galaxies, including our own Milky Way, glow with the light emitted when cosmic rays collide with interstellar matter within these galaxies. Galaxies shine so much more brightly in gamma ray light than the vast expanses of space between them that we can conclude that cosmic rays are far more common within, rather than outside, galaxies.

If, then, cosmic rays originate primarily from within each galaxy— and thus from within our Milky Way for cosmic rays detected on Earth—what could give them their tremendous energy? The ones that manage to penetrate the Earth's magnetic shield have energies that exceed 10,000 million times the energy of the molecules in our atmosphere. The rarest and fastest have astonishing energies—a billion times larger still.

Understanding the origin of the energy of cosmic rays lies, in part, with the knowledge of what they are made of. The particles that we can measure at ground level are mostly the products of a series of shattering collisions within the atmosphere above us, and, as such, cannot tell us directly about the original cosmic rays that created them. To get to the primary cosmic rays, instruments have to be lifted high into the atmosphere, or even above it, into space. Toward the end of the 1960s, instruments on high-altitude balloons had already given a pretty good idea of the composition of cosmic rays. State-of-the-art instrumentation flown in space today does an even better job in revealing what the population of primary cosmic rays is made of.

Cosmic rays in space are predominantly protons, the nuclei of hydrogen atoms. The next most abundant component is formed by alpha particles, which are the cores of helium, stripped of their electrons. The other components are hundreds of times less abundant than the protons. The curious thing is that all of the heavier, naturally occurring elements exist in the cosmic rays, and that the overall chemical mix is about the same as that of the solar system and of the local universe around us, which in turn is related to the composition of the human body. Apparently, cosmic rays are largely made up of the normal mix of elements out of which the stars themselves are made. There are some interesting differences in content, however, which tell us how they came to move at close to the speed of light. For example, whereas most elements in cosmic rays are represented in the same ratio as in the Sun to within about a factor of two, some are different by factors of up to a million. One such element that stands out in this way is beryllium, a very light element with only four protons in its core: the concentration of beryllium in cosmic rays is a million times higher than in solar gas. For boron, with five protons in its core, the cosmic-ray concentration is over 100,000 times higher. For lithium, with three protons in its core, the concentration is close to 2000 times higher. For helium and hydrogen the effect is inverted: they are somewhat less common in cosmic rays than in solar material.

These differences in the chemical mix of elements are the result of the collisions of cosmic rays with interstellar matter. In such collisions, the carbon and oxygen that are part of the cosmic ray population are broken up into smaller shards, which explains why there are so many more of the lighter elements—lithium, beryllium, and boron—in the cosmic rays near the Earth than in the Sun: these lighter cosmic ray particles are themselves natural byproducts of the original population. Similar effects enrich scandium, titanium, and vanadium in the cosmic rays by up to 1000 times by collisions of iron and nickel with interstellar matter. Understanding all this took decades of painstaking work, but the result is that we now know that original cosmic rays, before they collide with interstellar atoms, are largely made up of very nearly the same mix of elements as the gas throughout the Galaxy, with minor, comprehensible exceptions.

Having established that the original cosmic rays essentially have the standard galactic composition, we can rule out one possible mechanism that had been proposed to explain their very high velocities: they cannot be the direct result of stellar explosions. A star can exist only for as long as it has fuel to make it shine. When a heavy star runs out of fuel, it collapses under its own weight. Such a collapse causes so much energy to be released in such a short time by nuclear reactions deep inside it that the star explodes and becomes a supernova. The stellar remains in the supernova can shine as brightly as billions of stars together, at least for a few months. There would be enough energy in such an explosion to accelerate a fraction of its gases to near light speed, at which point they would be termed cosmic rays. However, at the end of their lives, stars have transformed almost all their hydrogen into heavier elements, including carbon, nitrogen, oxygen, iron, nickel, and more. If a stellar explosion were directly responsible for the acceleration of the cosmic rays, the cosmic-ray composition should be that of an old star, and therefore would be very different from the composition that we actually observe.

Ruling out that cosmic rays are a direct consequence of stellar explosions themselves, however, is not the same as ruling out supernovas as the ultimate cause of the high velocities of cosmic rays. Exploding stars cause shock waves to run through the interstellar gas. These are similar to the ones we hear when supersonic aircraft pass overhead, but they happen on a much bigger scale. Thereby, supernovas accelerate some of the interstellar gas atoms, thus making them part of the cosmic rays. However, although a stellar explosion is involved, the material that is being accelerated into cosmic rays is, in fact, largely the interstellar gas. One of the main pieces of evidence for that scenario is contained, like a fingerprint, within the cosmic rays themselves. In order to appreciate that, we need to look to radioactivity as a tool to prove the origin of cosmic rays as a side product of stellar explosions.

Radioactivity, the transmutation of one element into another by a nuclear process, comes in many guises. One of these is when an atomic nucleus captures one of its atom's own electrons. This bit of nuclear cannibalism converts one of the protons in the atomic nucleus into a neutron, while the disappearance of one constituent from the cloud of orbiting electrons changes the chemical properties of the newly formed atom. An isotope of nickel, nickel-59, can capture an orbiting electron to turn into an isotope of cobalt, cobalt-59. Under normal circumstances this would happen at a rate leaving half of the original atoms intact after a period of 76,000 years. So, the relative concentrations of the isotopes of nickel and cobalt within cosmic rays can tell us something about how long these atoms could have existed as plodding interstellar gas before being accelerated to near light speed in a supernova shockwave. That analysis revealed that the cobalt and nickel would have been part of slow-moving interstellar gas, with all or most of their orbiting electrons with them, for at least some 100,000 years before being stripped of their electrons and accelerated to become cosmic rays. Once stripped of its electrons, nickel-59 can no longer transition into cobalt-59 because there are no electrons around its nucleus to capture, so from that moment onward, the nickel and cobalt isotope abundances can no longer change, no matter how long they fly around the Galaxy as cosmic rays. That long initial time of floating around as interstellar gas means that it could not possibly be the exploding star itself that accelerated part of its own matter to become cosmic-ray material, but rather that such explosions power the gas floating in the interstellar medium around it to transform that into cosmic rays.

In summary, this, then, is the basic picture as we understand things: cosmic rays are atoms from the interstellar gas accelerated in areas of the Galaxy where heavy stars form and explode relatively close together, and then they scatter about the Galaxy for some 15 million years before leaking out. The random walk of the cosmic rays, scattering off changes in the galactic magnetic field as they fly around, causes them to approach the solar system equally from all directions: they come from no identifiable single event or from a particular direction, but instead have their origin in many supernova explosions all over the Galaxy, one of which goes off somewhere in the Milky Way approximately every 50 years. The cosmic rays that we detect on Earth and that create, among other things, the radioactive carbon-14 and chlorine-36 in our bodies, are part of a Galaxy-wide population of racing atoms that are the result of hundreds of thousands of exploding stars within the past 10 to 20 million years. That of course sounds like an eternity to us human beings, but it is merely 0.1% of the age of the cosmos, only one-quarter of the time since the dinosaurs became extinct, and going back to a time when mammoths, giraffes, and bears were already walking the Earth.

Unless there is a relatively nearby supernova, one that would out-shine all other stars in the sky, the galactic cosmic rays should be so om-nipresent, so diffuse, and so mixed up from travels around the Galaxy over millions of years, that the cosmic-ray intensity should not vary over time scales of days to years, or even centuries. And yet: the cosmic-ray intensity varies on all these time scales, by a larger factor on short time scales, while significantly less on time scales of years and more. Thus, several other things must be going on that contribute to the galactic cosmic-ray background and that affect how these rays get to the Earth.

The discovery of the change in the intensity of cosmic rays with time occurred in the late 1930s, mainly with the work of Scott Forbush, who observed patterns in the intensities of cosmic rays. He found that to de-scribe the slight daily variation in the cosmic-ray intensity, it was more im-portant to know what side of the Earth faced the Sun than what side faced a particular direction in the Galaxy (which differ from each other over the course of the year as the Earth orbits the Sun). He also demonstrated that it is important what side of the Sun faces the Earth: the Sun points more or less the same face to the Earth once every 27 days. That 27-day period stood out clearly in diagrams of cosmic-ray intensities, as well as in diagrams of the variations in the Earth's magnetic field, called geomagnetic storms. We now know that both of those variations have the same cause, namely the changing magnetic field of the rotating Sun as that field sweeps past all the planets and beyond, to the very edge of the solar system.

In the 1940s and early 1950s, the relationship between these phenom-ena and the nature of the Sun's interplanetary magnetic field were not known, however. When World War II came, Forbush's work was in-terrupted. After the end of that war, he looked at his data again and discovered two things at the same time. He noticed that there were short-lived increases in the cosmic-ray intensity following some of the largest explosions observed to occur on the Sun, namely two in 1942 and one after the war in 1946, when Forbush had resumed measur-ing cosmic rays. These increases of at most a few hours were followed by decreases in cosmic-ray intensities that lasted up to a few days. In those days, prior to space-based measurements by specially equipped satellites, the linkage between these phenomena could not be directly demonstrated: they were patterns in time, but there was no physical understanding that could provide an explanation. Not until the 1970s was it demonstrated that, in association with large explosions on the Sun, big clouds of hot, magnetized gases are expelled into interplanet-ary space. Whenever these clouds pass by the vicinity of the Earth, the pathways by which cosmic rays can reach the Earth are modified as they are deflected by the magnetic field within such clouds. That changes the cosmic-ray intensities for up to a few days, until the interplanetary storms travel onward, far beyond the Earth's orbit.

The cosmic-ray intensity also goes up and down as the number of sunspots goes down and up; that is, there is an inverse correlation with the Sun's magnetic activity. Through additional discoveries it became evident that the Sun, the Earth, interplanetary magnetism and cosmic rays were all connected, as was later confirmed by measurements with satellites. That work enables us to use carbon-14 and other radioactive particles to determine solar variability over the past thousands of years and to understand the causes of carbon-14 variations in all (living and dead) organisms on the Earth. To appreciate all that, however, we need to tell the story of the solar wind.

Key Points: Chapter 10

- Cosmic rays are nuclei from the interstellar gas, accelerated by shock-waves in areas of the Galaxy where heavy stars explode and become supernovas. These original cosmic rays are largely made up of the same mix of elements as the gas that is present throughout the Galaxy.

- Cosmic rays are influenced by collisions with the particles in interstellar space, thus changing the balance of their elements. They are also deflected by magnetic fields in the Galaxy.

- As a result of fluctuations in the galactic magnetic field, cosmic-ray particles randomly walk about in the Milky Way, remaining trapped inside it for, on average, 15 million years. Their erratic journeys after leaving distant origins cause them to scatter into our solar system equally from all directions.

- Each cosmic-ray particle entering the Earth's atmosphere is involved in a chain of collisions, causing a cascade of up to billions of particles, including protons, neutrons, electrons, and fragments of heavier elements. Cosmic rays have such strong penetrating power because they move incredibly fast, approaching the speed of light.

- **Cosmic rays are accelerated by exploding stars and are composed of ultra-fast atomic nuclei that tear through space at near light speed and that create a radioactive background in the elements in the human body.**

Kepler [1571–1630, long before the discovery of radiation pressure] ascribes the ascent of the tails of comets to the atmospheres of their heads, and their direction towards the parts opposite to the sun to the action of the rays of light carrying along with them the matter of the comet's tails.

ISAAC NEWTON (1642–1727)
in *The mathematical principles of natural philosophy (1687),*
Book 3, translated by Andrew Motte (1848)

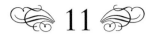

11

Tails in the Wind

Far beyond the outermost planet of our solar system, as far as halfway to the next nearest star, a vast swarm of chunks of primordial material has been orbiting the Sun since the beginning of the solar system itself. These chunks can be dust particles as small as a micron across or they can be multiple miles in diameter, all with highly irregular shapes. When these chunks are nudged from the paths they have been traveling on for billions of years, some small fraction may find themselves falling into the interior domain of our solar system, into the realm of its planets. Such orbital changes may occur whenever a star happens to pass relatively close to our Sun: once every 100 million years or so, a star approaches the Sun to within a few hundred thousand times the distance from the Sun to the Earth, which is close enough to influence the orbits of a multitude of these frozen dust bunnies. However, not only the effects of single stars journeying past the solar system are important: the combined effect of many more distant stars in the solar neighborhood of the Galaxy actually is somewhat larger than that of single stars passing by fairly closely. These gravitational forces act on the solar system as it travels around the Milky Way. The Galaxy may have vastly more stars than the Sun has planets, but their basic movements are the same: all stars swirl about the center of the Galaxy, subject to the collective gravitational pull of all stars within it, just as the planets orbit the Sun, where most of the mass of the solar system resides. Yet, there are slight perturbations in these orbits: similar to the slight perturbation of the Earth's orbit over time by the pull of the major planets (in the cycles discovered by Milankovitch, discussed in Chapter 7), so are the bodies in the outer solar system perturbed in their orbits around the Sun by the combined forces of stars relatively nearby.

The collection of oddly shaped rocky objects at the periphery of our solar system is called the Oort cloud, after the Dutch astronomer Jan Oort, who first proposed its existence in the middle of the twentieth century. The name Oort cloud does not convey what it really is: not a cloud like we have in our skies, but rather a very loose swarm that may comprise over 100 billion rocks. Obviously, no one has ever counted them, nor could we: as we look at smaller and smaller clumps, there are ever more of them. The bigger ones, however, can create quite a spectacle when they are pulled out of their orbit by gravitational perturbations and move into the solar system, first as comets in the solar wind and then possibly as meteors, should they descend into a planet's atmosphere.

At first, the falling object will sense little of the existence of the solar system other than the gravitational pull of the Sun. From its initial distance, the planets are not discernible, and even the Sun itself will be but a dot in the sky, only somewhat brighter than the other stars: seen from 10,000 to 100,000 times further away than we see the Sun from Earth, it is 100 million to 10 billion times fainter than it appears to us from the Earth. Most of these unsettled objects will not make it too close to the Sun: starting their fall from tens of thousands of Sun–Earth distances away, the inner solar system is a very tiny target that very few will be on a direct path to. Most of their orbits will be only mildly perturbed by passing stars and variations in galactic gravitation, so that by a very large majority they approach to only some hundreds or even a few thousands of Sun–Earth distances before their orbits take them further away again on their long elliptical paths around the faint, distant Sun.

For those objects from the Oort cloud that do approach the Sun to within just a few dozen Sun–Earth distances or less, the Sun will significantly warm their surfaces. Once warm enough, volatile compounds sublimate: the frozen gases transition directly from a solid to a vapor because there is no milieu for liquids on the surfaces of bodies in the vacuum of space (although some liquids may exist inside such bodies if they get warm enough). As the gases sublimate layer by layer from the surface, minuscule dust particles that lie frozen within them are released. The expanding cloud of gas and dust scatters the Sun's light, forming a faintly glowing cloud around the solid nucleus that astronomers refer to as the coma (from the ancient Greek word for hair). As the dust falls away behind the coma into a stream of particles, it scatters the Sun's light to create the appearance of a tail behind the coma: thus, a comet has formed.

Why should the dust fall away from the comet's nucleus? In a vacuum, one would expect no force other than gravity, which works the same for dust particles as for entire planets. So, if that were the only force acting, the comet's nucleus, its sublimating gases, and the embedded dust within them would all continue to travel along the same trajectory, merely creating a growing spherical fog around the nucleus as the gas in the coma expands. In order for the dust to move away from the comet's nucleus in a particular direction, there must be at least one additional force at work that acts differently on objects of dissimilar size or mass. There are indeed other forces at work. Not one, but in fact two forces push and pull on comet materials, in addition to gravity. Because these forces work differently on disparate particles, many comets have not just one but two tails. Depending on our perspective of the comet, these tails may be aligned, point in opposite directions, or have some intermediate angle between them.

One of the two comet tails—generally the brightest, if not the only one bright enough to be seen—is formed because sunlight scatters off the dust. Light not only carries energy, it also carries momentum, which means that if it collides with matter it pushes, or, in physical terms, exerts a repulsive force. This is true regardless of where in the universe we measure the effect. For example, if we shine a flashlight onto an object that subsequently either absorbs or reflects that light, a force pushes that object away from the light source. On the Earth, the forces of friction and viscosity with ground and air are so much larger than the force of light that in everyday settings nothing moves just because we shine a light on an object. In space, where there is essentially nothing to create friction, the force applied by the Sun's light is strong enough to gently push the dust from the comet in a direction away from the Sun. For a comet moving toward the Sun, the force exerted by sunlight causes the dust to lag behind the comet's nucleus and to drift slowly away from its original orbit into a wider one. In effect, the force of light acts such that the dust feels an effectively lower gravitational pull from the Sun. The main body of the comet is also subjected to a repulsive force, but it is far heavier and thus far more difficult to nudge off course.

As long as a comet is falling toward the Sun, the force of light pushes the tail in an arc behind it when looking in the direction of motion of the comet. Once the comet passes its point of closest approach to the Sun in its orbit and begins to move away from the Sun again, the radiation pressure pushes the dust away from the Sun slightly faster than the nucleus is moving. Consequently, the tail arcs ahead of the nucleus in its orbital segment moving away from the Sun.

A second comet tail is formed by the gases escaping from the comet's nucleus rather than by the dust. Many of the gas molecules lose one or more of their electrons and by doing so become electrically charged ions. These glow in a color specific to the comet's composition, often green or blue, rather than white as is the Sun's light scattering off the dust tail. The ions that make up the colored tail are exposed to another force in addition to gravity and photon pressure. Astronomers noticed that the second tail points away from the Sun; not in a moderate arc as does the dust tail, but in an essentially straight line very nearly directly away from the Sun.

What it is that creates this second, straight, colored tail always pointing directly away from the Sun was the subject of a big debate in the 1950s. Among the participants in this debate was Ludwig Biermann, a German astrophysicist then working in Göttingen. He was the first to point out that many comets had such a straight tail, and that these tails persisted over months as the comets traveled through their orbits around the Sun. Consequently, whatever was causing such tails had to persist at least that long. He suggested that these straight tails were caused by a sustained "corpuscular radiation", that is, by a continuous stream of particles moving away from the Sun. Within seven years of Biermann's inference, Gene Parker, working in Chicago, developed a theory that predicted that not only were particles indeed leaving the Sun all the time, but that they would be moving very fast, flying away from the Sun supersonically. He called this flow of several hundred kilometers or miles per second the solar wind.

Parker's ideas were so revolutionary that it took quite an effort to convince his colleagues of their correctness. In order to have a scientific study published in the professional literature, journal editors first review the quality of the work. Because of the often highly specialized nature of the manuscripts that they receive, the editors consult one or more experts working on topics similar to that of the manuscript, asking them for their opinions. The reviewers who commented on Parker's paper were not favorably disposed to his work at all. In due course, the editor Subrahmanyan Chandrasekhar decided to overrule the reviewers and he ignored their recommendation not to publish this work. After warning Parker of the consequences of potentially being wrong, he instructed the staff of the *Astrophysical Journal* to publish the young man's study.

Gene Parker turned out to be right. This was first demonstrated during a brief traverse from Earth to the Moon by the Russian Luna 1 spacecraft, the first spacecraft to travel past the Moon. That simple vehicle—an orb with antennae sticking out in all directions—took measurements during its 34-hour journey from the Earth to the Moon in January 1959, spending part of that time in the solar wind. The first segment of that trip was spent inside the Earth's magnetic cocoon, a volume of space into which the solar wind cannot penetrate because it is blocked by the terrestrial magnetic field. Once Luna 1 exited that domain it encountered the solar wind, but because the time during which it was measuring the solar wind over the rest of its active mission was limited, it remained unclear whether it had sensed a rare gust of wind or whether it had discovered a persistent and pervasive solar wind.

Unambiguous proof of the existence and persistent nature of the solar wind came in 1962, when the American Mariner 2 rocket was the first spacecraft sent out to encounter another planet, Venus. During its flight it obtained 113 days of measurements of the solar wind, making one full set of measurements every few minutes. Its records proved beyond a shadow of a doubt that the solar wind, although generally variable, exists as a sustained outflow from the Sun, much as computed from the basic equations of physics by Gene Parker.

Comets encounter the solar wind wherever they are within the solar system to some distance beyond the orbit of Neptune, the outermost planet. But the solar wind does not reach nearly as far out as the Oort cloud. At some intermediate distance from the Sun, the solar wind collides with the very tenuous gas that makes up the interstellar medium. Where the two are of comparable pressure, the solar wind slows down subject to the pressure of the interstellar gas. The two Voyager spacecraft, launched in 1977, have been traveling in the solar wind on their way to the outer planets and beyond. In late 2004, Voyager 1 passed the point where the solar wind slows enough that it is no longer moving supersonically and Voyager 2 did the same in late 2007. By the end of 2010, Voyager 1 had passed the point where the solar wind had stalled completely in the direction of its travel. The spacecraft was then over 100 times further away from the Sun than the Earth is. Although that means that the solar wind blows a very large bubble into the interstellar medium, the objects floating around in the Oort cloud spend most of their existence 100 to 1000 times further away than to where the solar wind reaches.

The collision of the solar wind particles with the atoms in the comet coma results in ionization of the gas, and the ions are subsequently dragged away from the coma. For comets within the part of the solar system where the planets are, the movement of the comet's nucleus relative to the Sun is much slower than that of the outflowing solar wind almost everywhere. The result is that the gas being blown out of the comet forms into a straight tail which points close to directly away from the Sun.

The main force causing this drag is the magnetic field that is carried within the solar wind. The ions that make up the solar wind originate from within the Sun's atmosphere, which is threaded throughout by a magnetic field. When these ions move outward from the Sun to form the solar wind, they drag a magnetic field with them. Once the atoms in the comet's coma become ionized by the collisions with the particles in the solar wind, they become locked to the magnetic field in the wind, and they are thus dragged away with the wind at speeds of about 300 to 700 kilometers (or 200 to 450 miles) per second.

The solar wind flows by every object in the solar system, not merely by comets, of course. In the case of planets that are poorly shielded by only a weak magnetic field, such as Venus and Mars, the solar wind can reach the outermost layers of the planetary atmospheres. Its impact gradually strips off atmospheric layers, blowing them away mixed in with the solar wind on a path out of the solar system. As a result, Mars—which probably started out with a thinner atmosphere than Venus in the first place—has been stripped of most of its atmosphere. Venus still has a thick atmosphere, although it did lose most of its water molecules as water vapor high in its atmosphere was gradually stripped away. This was caused by the combination of solar ultraviolet light breaking up water molecules and the solar wind subsequently stripping at least the hydrogen atoms away. The Earth, in contrast, is shielded by a magnetic field that is very much stronger than the remnant fields of Venus and Mars. The terrestrial magnetic field makes it virtually impossible for the solar wind to interact effectively with the Earth's atmosphere, and consequently the Earth's atmosphere is being lost to space extremely slowly.

The inner solar system is a hazardous place for comets, for several reasons. The closer a comet comes to the Sun, the faster its outer layers sublimate under the intense solar heat. As a result, every passage of a comet around the Sun leaves it with less mass to return with the next time, provided that it had enough mass to survive the previous passage in the first place. As it loses mass, it often happens that the comet simply fragments into perhaps a few, perhaps a multitude of chunks as the ice that holds the pieces together evaporates. These fragments subsequently all behave as comets in their own right, as they slowly drift apart on slightly different orbits. If the parent comet was large enough, this could result in a string of offspring comets, spread out along the original orbit, sometimes coasting with years or decades between successive fragments on their path around the Sun.

One such string of comets is particularly unusual, because the orbit brings them very close to the Sun. The so-called Kreutz group comets, named after the discoverer of their grouping, are what are known as Sun-grazing comets: they come to within a tenth of the solar diameter of the Sun's surface. The result is that most do not survive the encounter and evaporate well before even getting into the Sun's atmosphere. In 2011, the first observations were made of two Kreutz comets that did survive the approach to the Sun, at least for a while. The first did not survive the entire passage, but did glow while on its final descent through the Sun's atmosphere. Although it was not seen against the bright visible light of the Sun, it glowed brightly enough in ultraviolet light, in which the Sun appears as a very much fainter object, to show up in modern space-based detectors. Its mass of some 100,000 metric tons—as heavy as a mid-sized aircraft carrier—mostly evaporated within the last half hour of its descent. The second Sun-grazer seen within the Sun's outermost atmosphere did survive its close encounter with the Sun, but only briefly: within 16 hours of closest approach, its signal petered out as the last of its frozen material turned into vapor and dust.

These two Sun-grazing comets were particularly interesting because their ion tails were deposited into a part of the Sun's outer atmosphere that ordinarily cannot be observed because the light emitted from it is too faint to be seen next to the bright Sun. The observations of the behavior of the ion tails taught astronomers about the details of the magnetic field high above the Sun, and about how the solar wind that blows the tails out far into the solar system begins its journey from the Sun.

The threat of evaporating out of existence is shared by all snowballs bathed in sunlight. The planets form another hazard for comets. The big, heavy planets in the outer solar system—Jupiter and Saturn, in particular—can capture comets in their gravitational field and cause them eventually to collide with the planet. One such event was recorded by a series of observatories, including the Hubble Space Telescope, in 1994, when comet Shoemaker–Levy 9 (named after its discoverers Gene and Carolyn Shoemaker and David Levy) crashed into Jupiter. This was a very unusual event, not only because no other major impact had been observed before, but because it was in fact predicted. SL9, as it is known for short, did not hit Jupiter as it fell in from the outer solar system, but had actually been captured in the planet's gravitational field up to three decades earlier. Its orbit repeatedly brought it close to the top of Jupiter's atmosphere. During these close approaches, the gravitational forces pulled apart the core of the parent comet, forming a string of at least 21 new comets, each with a size of hundreds of meters (or yards) to two kilometers (or about a mile and a half) across. That train of comets fell into Jupiter's atmosphere in July 1994, leaving dark marks in the cloud cover, some of which lasted for several months.

If the orbits of a comet and of the major planet that it encounters are aligned in the right way, the moving planet's gravitational pull can slingshot comets out of the inner solar system, back to the Oort cloud where they came from. Such planetary slingshot movements were used intentionally to allow the Voyager 1 and 2 spacecraft to journey from one major planet to another over the course of several decades. In due time, the Voyagers were flung to the periphery of the solar system by their planetary encounters. They are now on their way out of the solar system and heading toward the Oort cloud. At the rate at which they are traveling they will not reach the inner edge of the Oort cloud until another few thousand years have passed: the Voyagers may get there, eventually, but they will long since have ceased to transmit data to Earth, and perhaps mankind itself will have forgotten about their existence.

In addition to the major planets, there are four much smaller Earth-like planets—Mercury, Venus, the Earth itself, and Mars—and their three moons (one for the Earth and two tiny ones, Phobos and Deimos, for Mars). If comets are small enough, and if the planet has an atmosphere, the friction with the atmospheric gases will heat up the comet's core to evaporate, if not to explode before colliding with the planet itself. The glowing trail following the nucleus of such a comet is called a meteor. When planets and meteors cross orbits and collide, there can be major consequences. If the original comet core is large enough, it can survive its travels through the atmosphere; smaller comets fragment either because of the gravitational pull they experience before they enter the atmosphere, or during their descent through the atmosphere.

If anything is left to hit the planetary surface, it is called a meteorite. Meteorite impacts can be devastating events that cause great damage to the surfaces of planets and their moons. The Earth's Moon is a good example of this: impact craters of all sizes can be easily seen all over its surface with just a pair of binoculars. Impact craters exist on Earth, too, but our oceans, biosphere, and the movement of the continents have wiped out most and hide others from our view. Some 40 craters larger than about 20 kilometers (or 12.5 miles) across can still be found on the Earth. The largest of these—the Vredefort crater, located in South Africa—has a diameter of about 300 kilometers (or 190 miles) and appears to be some two billion years old. The youngest of these large craters, a 52-kilometer (or 32-mile) hole punched into the Pamir Mountains in Tajikistan, is 25 million years old. Large impacts clearly did occur on Earth, but fortunately not too often. Their impact can be devastating, though: it is thought that the impact of one or more chunks of interplanetary matter into the Chicxulub area of Yucatan, Mexico, wiped out the last of the large dinosaurs on land and in the seas.

Between infrequent major impacts, material from small comets falls onto the Earth continually. It is estimated that on average, the Earth gains three pounds per second by capturing comet dust and tiny meteorites. In order to study what these small bodies are made of, specially designed experiments are needed because they do not make it to the ground intact. We cannot really learn the full story of their chemical makeup by looking at their larger counterparts either, because the meteorites that have been recovered from their impact sites bear little resemblance to their original form: most of the volatiles have evaporated during descent and impact, and much of their solids have melted together either by the heat of the friction with the atmosphere or by the heat released on impact.

Observations of comets through a handful of fly-bys by satellites (visiting comets Hartley 2, Halley, Borrelly, and Wild 2), and even an intentional crash into Tempel 1 to study the materials released by the impact, have revealed much of the properties of comet cores. The Stardust mission even returned some of the dust around comet Tempel 1 to Earth in 2006. Very-high-altitude airplanes (modified U-2 spy planes developed during the Cold War, now called Earth Resources 2, or ER-2) have also been used to scoop up comet dust from above elevations of 20 kilometers (12 miles). Comet nuclei are fluffy, porous clumps of volatile compounds—water or ice, nitrogen, carbon dioxide, ammonia, and other substances that we know as liquids or gases on Earth—mixed with non-volatile compounds, including a variety of silicates, some even in crystalline form, such as olivine and pyroxene, mixed with organic compounds and some rust. Apart from hydrogen, the lighter elements, and the noble gases, the composition of the comet cores is remarkably like that of the Sun. If the comets are basically made of the same mix of material that makes the Sun and its planets, then we can conclude that they all formed from the same cloud of interstellar material: comets are frozen bits of history, dating back to the formation of the solar system, some 4.6 billion years ago.

The material of comets, large and small, that comes down to Earth is mixed with the materials already here, and then used by all living organisms in our biosphere to build and rebuild their cells. We discussed in Chapter 8 how there are about a dozen carbon-14 atoms in every cell of the body, which were formed by collisions between cosmic rays and the Earth's atmospheric compounds sometime within the past 10,000 years or so. At the rate at which comet materials add to the Earth's carbon content, close to a million times as many carbon atoms in our bodies descended onto the Earth as meteors (or dust thereof) than are created from cosmic rays in any given period. At a globally mean human age of about 35 years, this implies that, on average, each cell in a person's body has a few tens of thousands of carbon atoms in it that were not on the Earth when that individual was born, including some that fell to Earth perhaps only weeks ago.

Key Points: Chapter 11

- Slight orbit perturbations can cause clumps of galactic matter to enter the inner solar system, becoming comets in the solar wind, meteors when they descend into a planet's atmosphere, and meteorites upon impact with the planet's surface.

- In addition to being pulled by gravity, a comet's core and the dusty cloud around it are pushed by pressure from sunlight, and together they create a bent comet tail pointing away from the Sun, in an arc. The solar wind causes the appearance of a second, straight, colored comet tail that points close to directly away from the Sun.

- The ions of the solar wind originate from within the Sun's atmosphere. Atoms around the comet nucleus become ionized by the Sun's light and are then dragged away with the magnetic field that accompanies the wind.

- Material from small comets falls onto the Earth continually. On average, the Earth captures about three pounds of comet dust and tiny meteorites per second.

- Each cell in a human body of average age contains several tens of thousands of carbon atoms from comets that fell to the Earth within mere decades.

- **Comets are frozen bits of history, dating back to well before the formation of the solar system, some 4.6 billion years ago. They are basically made of the same materials that make the Sun and its planets, and were formed from the same cloud of interstellar material.**

Even
After
All this time
The Sun never says to the Earth,
"You owe me."
Look
What happens
With a love like that,
It lights the whole sky.

HAFIZ (c.1325–89),
in *The Gift: Poems by Hafiz, the Great Sufi Master* (1999)

12

A Magnetic Heartbeat

The solar wind is never steady. Over days to weeks it changes in speed over a range from about 300 to 700 kilometers per second (or about 200 to 450 miles per second). In sudden gusts it can reach over 2000 kilometers per second (or about 1200 miles per second), moving so fast that it would cross the entire continental United States, Europe, or Brazil in a matter of 2 seconds. Over the years, the average wind speed and the frequency and strength of its gusts vary gradually, modulating the weather in space over an 11-year cycle, which in turn varies with a slower, irregular beat over many decades. That slow beat in time is found imprinted in the concentrations of carbon-14 and other radioactive atoms on Earth, including those in our bodies. The connection between those radioactive atoms within us, the cosmic rays that create them, and the solar wind reaches into a magnetic dynamo deep within the Sun. Some of the effects of that solar dynamo were known long before we understood their magnetic cause. The longest-known signature consequences of the Sun's magnetic field are the northern lights and the patterns in the sky around a total eclipse.

An eclipse of the Sun has always made deep impressions. In modern times, we understand this phenomenon as merely a predictable alignment of the Earth, the Moon, and the Sun. During an eclipse, the Moon blocks out all or part of the Sun's light over a narrow swath across the Earth. In times long past, the phenomenon was viewed with great awe and fear. Among the oldest records found to date is a Chinese description of a solar eclipse that occurred over 4000 years ago. In those days, it was thought that a gigantic, yet invisible dragon swallowed the Sun in its entirety. It was believed that raising great alarm on Earth could somehow undo this cosmic gluttony and indeed the Sun reappeared in the sky, as it would have, of course, without any such activity. Ancient solar eclipses are reported to have stopped battles or to have altered the outcomes of wars. Even the Bible mentions a few eclipses, possibly including one historical one that was recorded in ancient Assyria. Some predictive capabilities for the occurrence of solar eclipses, developed using patterns found in observations of past events, were established as early as 4300 years ago, and further advanced by Greek astronomers over 2000 years ago. These methods were much refined after Newton's discovery of the law of gravitation (first published in 1687), which enabled the accurate computation of the orbits of the Earth around the Sun, and of the Moon around the Earth.

During a total solar eclipse, which occurs because the Earth–Moon distance is just right for the apparent sizes of the Moon and Sun in the sky to be very nearly the same, the atmosphere of the Sun reveals itself. This atmosphere is otherwise invisible to the unaided eye because of the overwhelming brightness of the solar surface. When viewed during a total eclipse, the innermost visible structures in the solar atmosphere, made up of clouds of dense gas that are known as prominences, glow red in a layer called the chromosphere. Around and above these is an omnipresent but highly structured white glow that extends out to several times the solar diameter. This outermost part of the solar atmosphere is known as the corona. Within this glow, we see a variety of formations that have been described as rays, helmets, cusps, and—for as long as historical records go back—wings. We now know that these faint, wing-like structures are shaped by the Sun's magnetic field, which is stretched out by the flow of the surrounding solar wind. In centuries past, however, nothing near that understanding existed.

Walter Maunder, who wrote about this in a book published in 1908, explained that these coronal structures seen during eclipses are likely the reason why so many ancient cultures had Sun symbols showing wings extending from a golden disk. Maunder wrote: "It seems exceedingly probable that the symbol of the ring with wings owed its origin not to any supposed analogy between the ring and the wings and the divine attributes of eternity and power, but to the revelations of a total eclipse with a corona of minimum type". For a long time, it remained unclear whether the "wings", "feathers", and rays seen during solar eclipses were structures associated with the Sun, or were a signature of the Moon's atmosphere, or were perhaps optical manifestations originating within the Earth's atmosphere. Over the years, with increasing observational capabilities and understanding of the universe (including the fact that the Moon has no atmosphere), it became clear that the structures seen around the Sun during total eclipses were truly solar phenomena.

The solar corona has been imaged during more than 100 total solar eclipses that have occurred since the development of photography enabled astronomers to take pictures of the glowing chromosphere and corona against the darkened sky. The first eclipse photograph dates back to the total eclipse of July 28, 1851, and was taken in Königsberg (now known as Kaliningrad, a Russian exclave nestled between Poland and Lithuania). The earliest photographs generally show only the chromosphere correctly exposed, but from no later than 1869 photographs imaged the rays of the corona as well. These structures change so markedly from eclipse to eclipse that it is clear they reveal a continually changing atmospheric framework, and not merely the same structures viewed from different angles as the Earth's perspective on the corona changes over time.

Another solar phenomenon turns out to be closely related to the solar corona: sunspots. These dark patches on the Sun are hardly ever large enough to be visible to the naked eye, even though many are larger than the Earth; the Sun is just too far away for us to easily see the relatively small, yet intrinsically sizable blemishes known as sunspots. Even those large enough to be visible to the naked eye in principle can be viewed only around sunrise or sunset, when the sunlight is so attenuated in the Earth's atmosphere that we can hazard a brief look at the glowing disk. The oldest records of such observations date back some 2400 years, when they were entered into Chinese archives. Very occasional mentions of sunspots also appear in European records—from the third century BCE in Greece onward—but these single sightings were generally interpreted as transits of hypothetical planets in their orbits between the Sun and the Earth. It was not until 1610 and 1611, when Johannes and David Fabricius, Thomas Harriott, Galileo Galilei, and Christoph Scheiner all pointed their own versions of a very recent invention at the Sun, that sunspots were recognized as a solar phenomenon. That invention was the spyglass, the precursor of the modern-day telescope. This device, claimed by the Dutch spectacle maker Hans Lippershey in 1608, enabled magnified viewing. Lippershey's own "perspective glass" had only a threefold magnification. Harriott's later version increased that to six times. Galileo's version had a magnification of 26 and combined that with an aperture that collected almost 60 times as much light as the open pupil of the human eye.

Galileo pointed his telescope at a variety of objects in the sky. The superior light-gathering and magnifying power of his instrument enabled his discovery of the four brightest moons orbiting the planet Jupiter. This was one of the observations that prompted him to propose that— as an enlarged counterpart of the Jovian system with moons orbiting the planet—the Sun was at the center of the universe, and around it all objects orbited, rather than around the Earth. That proposition, however, landed him in severe trouble with the Church, for which he nearly paid with his life (and he would have been wrong anyway: it is now thought that the universe is infinite and has no center, but he was right insofar as the planets in the solar system orbit a common center of mass that is not far from the center of the Sun). Projecting the Sun's image onto sheets of paper, Galileo created series of detailed drawings (digital copies of which can be readily found on the internet) of sunspots on the solar disk. Stringing these frames together into a movie, we can easily come to the same conclusions as Galileo: sunspots are structures that evolve in a matter of days, even as their displacements in the drawings from day to day reveal that the Sun itself rotates roughly once every 27 days as viewed from the orbiting Earth. Galileo observed that whenever sunspots were close to the edge of the solar disk, they were foreshortened, from which he argued that they were indeed on the solar sphere itself and not unknown objects moving in front of an unblemished Sun.

In the subsequent four centuries, sunspots were carefully measured at multiple observatories, a practice that continues to this day. By 1843, Heinrich Schwabe noticed the gradual rise and fall in the number of visible sunspots on the Sun in a 17-year record. Rudolf Wolf then looked through the existing sunspot records and reconstructed an overall view going back to 1745. The compiled measurements clearly showed that sunspots consistently increase and drop in numbers on a roughly 11-year cycle. When Walter Maunder and his wife Annie were preparing for a 1904 meeting of the Royal Astronomical Society, they made another historical discovery: they plotted the positions of groups of sunspots in solar latitude as a function of time. Annie Maunder later described that process: "We made this diagram in a week of evenings, one dictating and the other ruling these little lines. We had to do it in a hurry because we wanted to get it before the [Royal Astronomical] Society at the same meeting as the other sunspot observers." That diagram was the first rendering of what every solar physicist nowadays knows as "the butterfly diagram".

The butterfly diagram clearly reflects why that name is the perfect description: at the start of each sunspot cycle, spots begin to emerge at mid-latitudes on both the northern and southern hemispheres, then successive generations of spots emerge in a wide band around mean latitudes that gradually migrate toward the equator, until these bands fade out at the end of each sunspot cycle. Hence, when drawing the latitudes of sunspots on the northern and southern solar hemispheres against time, the diagram completes two fairly symmetric wings that look like a butterfly's open wings as seen from above. During World War II, Annie Maunder, then living in London, gave the original diagram to a colleague, who took it to the United States to keep it safe from the German bombing raids during the Blitz. There it eventually made it to the High Altitude Observatory in Boulder, where it is still displayed as one of the major discoveries in astrophysics.

The sunspot cycle was discovered well before the nature of sunspots was known. Initially, sunspots were interpreted variously as clouds over the Sun, as openings in a cloud deck affording a view to the real solar surface, or as chunks of rock floating around in glowing lava. In 1908, George Ellery Hale, at the Mount Wilson Solar Observatory in southern California, demonstrated that sunspots are clusters of very strong magnetic fields. In order to do this, he used an effect discovered by Pieter Zeeman in 1896. Zeeman was awarded the Nobel Prize in physics for that discovery in 1902, but before that happened he was in fact fired because of his performing the experiment in the first place, going against the direct instructions of his superior! Zeeman had discovered that when light passes through a strong magnetic field, its color (characterized equivalently by a wavelength or by a frequency) splits into two very similar but slightly different ones, and when an instrument can unravel light adequately—as could Hale's telescope, completed in 1907—this so-called Zeeman effect can be used to demonstrate the existence of a magnetic field and to measure its strength. This method is still commonly applied, not only for the Sun, but also for other stars and objects much further away from the Earth. Continuing his studies of sunspots, Hale showed in 1914 that their magnetic polarity patterns reversed from one cycle to the next. Continuous magnetic records started in 1917 show that this is a fundamental property of the sunspot cycle. Because of this alternating polarity pattern, the processes that generate the Sun's magnetic field—still poorly known in their details—are now known collectively as the solar dynamo.

That dynamo turns out to have some surprises in store that were not initially recognized as having a solar origin. One long-known such phenomenon is that of the northern lights, the auroras borealis. Myths about their cause include reflections off the shields of the Valkyries (creatures who were supposed to decide which soldiers die or live in battle) or off schools of herring in the northern seas, and that they are the distant glow of divine fires. Other myths hold that they are the signatures of hunting gods or of snow brushed upward by the swooping tail of a magical fox, while still others maintain that the auroras could even take people away. During the eighteenth century, the realization emerged that another phenomenon, namely variations in the Earth's magnetic field, were somehow connected to the auroras. Magnetism had been known, although not understood, for a very long time, including the existence of some force on the Earth that always oriented a small floating or suspended magnetic element in the same direction. The application of that force for navigational purposes came about with the development of the compass. It appears that the Chinese already had functional directional devices over 4500 years ago, and that these were used, for example, to guide armies in the right direction through the worst of fogs. These may have been magnetic but perhaps they were mechanical devices that kept track of the turns of the wheels of chariots. Sometime between the third and fifth century, though, magnetized needles—either floating in a liquid or hanging by a thin thread—were developed by the Chinese and used for navigation. True shipboard compasses were in use by the twelfth century.

In Europe, the first useful compass for navigation on the open seas appears to originate with Peter Peregrinus de Maricourt, who wrote about the design of such a navigational tool in 1269, with not only a dry pivoting needle rather than a floating one, but with a design that included a stable housing with directional indicators, and with a sighting device to measure directions of objects relative to the direction of the compass needle. It is remarkable that it took over 350 years of compass use before someone remarked on the existence of low-level swings in compass pointings on time scales from minutes to days. In 1722, George Graham, a maker of clocks and astronomical instruments in London, observed that his compass needles not only showed a daily variation (of no more than about a sixth of one degree) but that, on rare occasions, they made irregular swings. These swings could be as large as half a degree within a minute, increasing to several degrees in a matter of hours, even with the compass housing lying perfectly still and with Graham taking great care to ensure that neither he nor his workplace were causing the disturbances. Others observed and reported a similar phenomenon in 1738. Among them was the Swede Anders Celsius, who remarked (in 1741) on the apparent simultaneous occurrence of the compass variations and auroras, confirming this with hundreds of similarly correlated pairings of magnetic disturbances and auroras. Alexander von Humboldt, working in Berlin in 1806, observed a strong geomagnetic storm using a compass. He became so intrigued with the phenomenon that he led the development of a worldwide web of geomagnetic observatories. From 1832 onward, the geomagnetic field has been continuously monitored and recorded in many places around the world, revealing both the small, irregular short-term swings and a very much slower gradual drift of the direction of the magnetic north pole.

The commercial use of telegraph systems that started in the mid-1830s brought into view another manifestation of the geomagnetic storm: the Earth's changing magnetic field induced high voltages and substantial electrical currents through the long metal wires used to carry messages. These voltage spikes were sometimes large enough to injure the telegraph operators or to cause sparks that could burn telegraph stations to the ground and start prairie fires. In those days, the telegraph network had a critical function in fast long-distance communication, and the interests in geomagnetic storms shifted from the esthetic aspects of the auroras to the pragmatic predicaments of entrepreneurs cut off from their markets, and to the concerns of army leaders isolated from their troops.

In 1859, Richard Carrington proposed the connection of terrestrial auroras with physical processes outside the Earth. Carrington, an astronomer by training but also part-time director of the family brewery, was observing the Sun on September 1, 1859, when he saw sudden brightenings in between well-developed sunspots. Carrington, being careful, noted that it was something that he "believe[d] to be exceedingly rare". Rare it was indeed: no one had ever seen anything like it until then. He could not find anyone to independently witness the event with him during the few minutes that it lasted. Fortunately, however, Richard Hodgson observed the same event from an observatory in Highgate, in the northern suburbs of London. Both Carrington and Hodgson reported their observations at a subsequent meeting of the Royal Astronomical Society. They had witnessed what we now call a solar flare: a sudden, major release of electromagnetic energy from the Sun's atmosphere that lasts for a mere few minutes. The flare they saw was unusually large, hardly ever surpassed in the one and a half centuries since. Its enormous burst of energy was equivalent to that of hundreds of billions of hydrogen bombs going off at the same time.

By 1895, much of the knowledge was in place to reveal the real connections between all these phenomena. It was the Norwegian Kristian Birkeland who first put enough of it together so that our understanding of the Sun–Earth connection could begin to mature. In his laboratory, Birkeland used a magnetized sphere (called a terrella, for "little Earth", first used to understand compass pointings by William Gilbert in 1600), which he placed in a vacuum to mimic the Earth and its magnetic field in space. He then directed cathode rays (charged particles now known to be electrons, but then only just discovered and still highly debated) at the terrella through the vacuum to mimic the solar wind blowing past the Earth, and saw that these rays caused the sphere to glow near both of its magnetic poles. He had thus artificially created something not dissimilar from the northern aurora borealis and the symmetrical phenomenon on the other pole, the southern lights, or aurora australis. The aurora australis had hardly ever been observed by then, because Antarctica was mostly yet to be explored, with Roald Amundsen and Robert Scott not reaching the south pole until sixteen years later, in 1911. Birkeland was very happy with his experiment. He hypothesized that sunspots emitted cathode rays and that these generated the auroras on impact with the Earth's atmosphere. He did not realize that he missed several critical steps in the chain of processes connecting sunspots, auroras, and magnetic storms, but it would take until the 1970s before the last of those were discovered. All of these steps became possible because of advances in instrumentation and, ultimately, because of access to space.

Today, we routinely make maps of the Sun's magnetic field, not only within the sunspots but anywhere on the side of the rotating Sun that happens to face the orbiting Earth. The Sun's magnetic field is composed of a hierarchy of north-next-to-south polarity regions that bob to the surface from deep inside the Sun. The largest, strongest of those regions contain sunspots. These are relatively dark and cool compared with the surrounding solar surface, which is called the photosphere. Smaller patches of magnetism surround the sunspots in great numbers, shining slightly more brightly than the sunspots; these are known by their old name of faculae, which derives from a Latin word for torches. Nothing in the Sun's magnetic landscape—gasscape would be more appropriate—is fixed. Sunspots come and go on time scales of days to weeks, while the surrounding faculae wander about and may last for similar times until they run into others, which can strengthen or weaken them, or even obliterate them altogether, the outcome of the interaction depending on whether the clashing faculae are of like or of opposite magnetic polarity. In the background of all sunspots and faculae, there is a weak but very large-scale bipolar field, which is much like the Earth's global magnetic bipole. The solar magnetic field strengthens and weakens every 11 years. Its polarity flips at the end of each such cycle, resulting in what is a 22-year solar magnetic cycle of alternating north–south polarity.

On scales much smaller than the sphere as a whole, the motions of the gas within the Sun force the patches of magnetic field to shift around on the solar surface. This in turn induces field changes elsewhere. All such changes are associated with electrical currents running through the Sun's interior and atmosphere. These electromagnetic patterns become stressed as they are moved about. The stressed magnetic field of the Sun causes the gas in the Sun's corona to be heated to millions of degrees, and so it glows in ultraviolet and X-ray wavelengths of light. In fact, the gas in the solar outer atmosphere is heated so much that neither gravity nor the Sun's magnetic field can keep it from flowing into interplanetary space. As the highest gas flows away to form the solar wind, it drags strands of magnetic field with it. This is what shapes the high corona at the base of the solar wind into the pointed tufts or streamers seen during eclipses. The density of the gas there is extremely low and, for example, far less than what we think of as the "vacuum" experienced by spacecraft in near-Earth orbits. Even at that very low density, however, there is enough to scatter a very small fraction of the Sun's surface light into a haze that is shaped by the magnetic field, revealing the shape of that field when the glare of the solar disk is blocked, such as during a total eclipse.

Much of the time, the X-ray glow and the solar wind are gradually modulated by the evolving patterns of the surface field and its sunspots. At times, however, the stresses in patches of strong magnetic field build up so strongly that major explosions occur: so much energy is converted from the magnetic field that the field cannot contain the explosion. Then, vast eruptions may cause parts of the solar atmosphere to expand into space in what are called coronal mass ejections. These mass ejections temporarily disrupt much of the Sun's atmospheric magnetic field. After such a major eruption it takes one or two days before the patterns of the magnetic field resume their quiescent evolving motions.

Eruptions of the Sun's magnetic field can occur in large areas of relatively weak magnetic field or in compact regions of strong field. If they occur in the latter, they are generally associated with a solar X-ray flare, but if in a weak field environment, the afterglow is fairly faint and occurs in the extreme ultraviolet, without extending into the X-ray domain. In that case, the flare is not recorded with conventional instrumentation. Flares in regions of strong field result from destabilizations of the magnetic field, but only the strongest of these are generally capable of breaching the overlying magnetic field and of escaping into interplanetary space to form a coronal mass ejection.

These ejections throw gases and their embedded magnetic fields away from the Sun at speeds of up to 2000 kilometers per second. More typical speeds range from 700 to 1200 kilometers per second. Even though that sounds like it is going really fast—and, of course, it is—the characteristic speed would move such an ejection by only 9% of the solar diameter during a two-minute solar eclipse. So even if observers of eclipses would happen to be looking just as a coronal mass ejection was developing, they would see the protruding solar ray tufts change only a little. This is the main reason why such ejections were unknown until special instruments developed in the twentieth century revealed their existence.

The advances in scientific instrumentation were important because coronal mass ejections turned out to be the link that connects the solar magnetic field to the strongest geomagnetic storms and auroras. Because coronal mass ejections are not necessarily associated with visible solar flares and most solar flares are not associated with coronal mass ejections, however, this connection remained obscure and controversial for a long time. By the mid-1970s, with X-ray observations being taken from space and with artificial eclipses being created by specialized ground-based observatories in "coronagraphs", a complex but complete picture finally emerged.

Most coronal mass ejections miss the Earth, but in many cases their much-diminished remnants do envelop the Earth on their way to the outskirts of the solar system. This happens when the ejection leaves the Sun from the vicinity of the center of the disk (as viewed from the Earth) and moves more or less radially from the Sun. Once such an ejection arrives near the Earth, the magnetic field within it and that of the

Earth interact. It is as a result of that interaction that atoms are accelerated that, when they hit our atmosphere, cause oxygen and nitrogen to glow to form the aurora. The details have now been worked out enough to understand why auroras occur preferentially at latitudes around the Arctic and Antarctic circles, and why auroras reach closer to the equator when storms are stronger, and to comprehend that the aurora-generating atoms do not come directly from the Sun. Not surprisingly, the Earth is not unique in having a planetary magnetic field: Jupiter, Saturn, Uranus, and Neptune also have fairly strong dipolar fields. Observations with, for example, the Hubble Space Telescope have revealed auroras on those planets, too, as expected from our current understanding of these phenomena.

Every day, several coronal mass ejections leave the Sun, more during years with many sunspots than in other years. Some are faster than others, and the slower mass ejections are frequently overtaken by and then merge with faster ones. All these eruptions occur on the background of the persistent yet also variable solar wind, for which periods of relatively slow streams alternate with periods of fast ones. The varying streams and gusts, and the patterns of alternating magnetic polarity in the solar wind make for a complex arrangement of the Sun's magnetic field as that is being stretched in the wind moving outward from the spinning star.

As it turns out, the solar wind and its multitude of embedded magnetic irregularities create a formidable barrier for cosmic rays that penetrate the solar system. That, in turn, leads to variations in the number of radioactive carbon-14 atoms found in every organism on the Earth. This is because once galactic cosmic rays come to within the distances reached by the solar wind—somewhat more than 100 times the Sun–Earth distance—they are subject to the forces of the solar wind's magnetism. The rays travel in big spiraling motions, swirling around the local direction of that field. The irregular variations of the field inside the solar wind turn the paths of cosmic rays into random walks on top of their spiraling motions. Consequently, it takes a long time for them to meander from the edge of the solar system to near the Sun and to where the Earth's orbit is located: it takes them several weeks to months before they penetrate as deeply as the Earth's orbit. In contrast, if they could fly in a straight line instead, their velocity of close to the speed of light would bring them from the edge of the solar system to the Earth in just over half a day.

Even as the cosmic rays are straying into the inner parts of the Sun's domain, the heliosphere, in their somewhat random trek, the solar wind blows them out again. The result of this inward migration subjected to an outward flow is that there is a very pronounced gradient in the typical cosmic-ray density: near the edge of the solar system the intensity of cosmic rays is high, but around the inner planets closest to the Sun that intensity is much lower, because the solar wind manages to keep most of the cosmic rays away from there. The fastest, most energetic cosmic rays are affected less by the solar wind and its magnetic field, so the ratio of cosmic-ray densities in the outer and in the inner heliosphere also depends on the speed of the cosmic rays. If we look at the most common cosmic rays that have enough energy (1 billion electronvolts) to cause nuclear reactions in the Earth's atmosphere, the intensity of galactic cosmic rays near the Earth is several hundred times less than in interstellar space, beyond the bubble blown by the solar wind. The decrease in intensity is most pronounced within the inner solar system because that is where the solar wind's magnetic field is strongest: at Neptune (at 30 Sun–Earth distances) the cosmic-ray intensity is only a little below that of the interstellar medium, at Jupiter (at five Sun–Earth distances) the difference is already substantial, and then the drop in intensity increases rapidly as cosmic rays approach the innermost planets of Mars, Earth, Venus, and Mercury.

The rates and strengths of coronal mass ejections vary with the patterns and strengths of the magnetic field on the surface of the Sun, as do the properties of the persistent solar wind. As a result, the cosmic-ray intensities at the Earth reflect the 11-year sunspot cycle, and indeed the full 22-year magnetic cycle that is made up of pairs of sunspot cycles with opposite magnetic polarity patterns. The modulation of cosmic rays by the irregularities of the solar wind is directly seen in the nuclear fragments created when cosmic rays hit our atmosphere. These fragments have been measured continuously since the early 1950s (in part a consequence of the fear of nuclear tests or attacks during which similar particles would be created). The neutron intensities at the Earth's surface, for example, go up and down substantially by several tens of percent over the years, with clear signatures that parallel the 22-year magnetic cycle of the Sun. These modulations are also seen in the radioactive atoms created higher in the atmosphere, such as carbon-14. Because each year only a small amount of new carbon-14 is created and then mixed in with the vast reservoir of cabon-14 atoms already present, however, it is essentially impossible to see a fingerprint of the 22-year solar cycle in carbon-14 measurements overall. The longer-term multidecade variations in the strengths of these cycles do show up, enabling us to map the solar magnetism back into prehistoric times.

Carbon-14 is not the only radionuclide created by cosmic rays within the Earth's atmosphere: among the radioactive particles that live long enough and that are created in sufficient quantities to be useful for scientific analysis is beryllium-10. That element is not absorbed into any chemical that is a significant part of the Earth's biosphere, but it does attach to aerosols that subsequently fall out of the atmosphere in precipitation. Where the precipitation is in the form of rain, beryllium-10 is mixed with the global water reservoir, and any temporal signal is thereby washed out as it mixes with all of the beryllium-10 already there. Where the precipitation is in the form of snow falling onto long-lasting ice fields, however, as is the case for glaciers in high mountains and in the polar regions, the concentration of beryllium-10 within the snow is preserved year by year, in layer upon layer. These layers reveal variations in concentration over thousands of years. It takes one to two years for the beryllium-10 to migrate from the high stratosphere, where most of it is produced, into snow layers; nonetheless, although the 11-year cycle is somewhat shifted and smoothed in time, it is mostly preserved.

The analysis of the beryllium-10 records over hundreds of years of fallen snow that has turned into ice in glaciers in Greenland and Antarctica not only shows the 11-year sunspot cycle, but also reflects substantial slower trends. These are a combination of the slow changes in the Earth's climate and the even slower ones in the Earth's magnetic field (the short swings in the Earth's magnetic field caused by the impact of the variable solar wind are too small and short-lived to matter). The climatic changes are part of the equation because they determine how the beryllium-10 is mixed throughout the atmosphere and where the bulk of that element precipitates, either as rain in the oceans or as snow on high-altitude or high-latitude permanent ice fields. The slow shifts in direction and strength of the Earth's magnetic field come into the picture because the galactic cosmic rays encounter that field on the last leg of their travels into the Earth's atmosphere: because these rays are sensitive to any magnetic field, and because that of the Earth is quite strong compared with that in the solar wind, changes in the Earth's magnetic field do show up in the concentrations of radioactive particles over the centuries and millennia.

The beryllium-10 record shows a modulation by a factor of approximately two over the past 60,000 years, with the strongest enhancement 40,000 years ago. That period is named the "Laschamp event", named after a French lava field in which a decrease in the Earth's magnetic field was first discovered. That was a time when the Earth's dipole field weakened for several centuries to only one-twentieth of the current strength: compass navigation would have been impossible in those days. But then, traveling was neither easy nor common anyway: that period was deep within the last ice age, at a time when people—modern *Homo sapiens*, with Neanderthals recently having become extinct—had slowly spread across Africa, the Middle East, Asia, and Australia, but with few traces of human presence found in Europe, and none having been seen in the Americas. Despite the Earth's weak global magnetic field during the Laschamp period, a rather patchy remnant field still shielded the Earth fairly well from cosmic rays, so that the carbon-14 count in humans has likely always been within a factor of approximately two of the present value.

The study of the long-term changes in the Earth's magnetic field, by the way, is aided by a combination of carbon-14 dating and a phenomenon referred to as the "Curie point". The latter brings us back to Pierre and Jacques Curie, who worked in the early days of discovery of radioactivity, cosmic rays, and magnetism. Pierre Curie discovered that a magnetized substance would lose all trace of its magnetization if heated above a certain point, and that if cooled it would "freeze" into its molecular patterns the direction and strength of the Earth's magnetic field at that location and time. This happens to cooling lava, too, which is how the low field strength of 40,000 years ago could be detected in the Laschamp lava field. In a related manner, geologists are helped by archeologists in studying the Earth's magnetic field over time: any place where a fire has been hot enough, and where remnants of the burned wood or coal are still present, would have information on when the material was heated and then cooled again through its carbon-14 content, while the geologists could then analyze the direction and strength of the Earth's magnetic field locked into the molecular patterns of the rocks on which the fire rested at that time.

The combination of different sciences needed to disentangle changes in the solar and terrestrial magnetic fields takes much detective work and perseverance. However, the collection of enough fireplaces spread across the Earth and time has helped us to determine the magnetic field of both the Sun and the Earth over millennia through the intermediary of galactic cosmic rays. Currently, after much research, carbon-14 dating that has been verified through tree-ring analyses extends back solidly to almost 13,000 years ago, with less accurate dating going back to 17,500 years before the present. Carbon-14 dating without explicit use of tree rings, such as the dating of ancient fireplaces, enables researchers to go back at least 50,000 years, although with uncertainties increasing from five years as far as 11,000 years ago, to up to 100 years when looking at materials from over 40,000 years ago.

The combined studies of radioactive particles and the analysis of the terrestrial magnetic field reveal that the Sun has been, on average, about as active as it is now for the last million years, and that it has changed relatively little even over the past billion years. However, these studies have also revealed periods during which multiple successive sunspot cycles are unusually strong. One of those we experienced during the past few decades, for example, during which sunspot numbers during cycle peaks reached values almost double of what was observed in the first half of the twentieth century. At the other extreme of solar activity are periods of sustained low sunspot numbers. One such interval started not long after sunspots were first discovered using the telescope: just 33 years after western European scientists—including Galileo Galilei—discovered the existence of sunspots the Sun went into a 70-year period with very few such spots. That so-called Maunder minimum had sunspot numbers that were, averaged over 11-year periods, over 20 times lower than in previous decades.

Radionuclide studies have given us glimpses of the Sun's activity over at least a dozen millennia, that is, to before the development of cities in ancient Mesopotamia, Assyria, and Egypt. As we do these days, the sky watchers of old would occasionally have seen the consequences of solar activity in the form of auroras. Without technologies sensitive to geomagnetic storms and lacking telescopes to observe the Sun, however, auroras would be all that they could see of the Sun–Earth connection. They certainly had no idea of the generation of carbon-14 following the struggle between the galactic cosmic rays entering the planetary system and the outward-moving solar wind, or of the fact that it could be used to date the artifacts from their cultures and others.

From the perspective of this story about the makeup of our bodies, it is interesting to realize that the carbon-14 dating technique is hardest for the most recent century. This is because humans have been interfering with the carbon-14 to carbon-12 mixing ratio in quite substantial ways since the beginning of the industrial revolution. One manner in which this happened was by the use of fossil fuels in the form of gas, coal, and oil. Both coal and oil have been used for centuries, of course, but the industrial revolution in the second half of the nineteenth century, followed by rapid growth in energy use throughout the twentieth century and into the new millennium, have put a lot of aged carbon back into the atmosphere in the form of carbon dioxide. In 2010, for example, an estimated 34 billion tons of carbon dioxide was released into the atmosphere by intentional burning of fossil fuels (not counting the effects of several vast and persistent underground fires in deep coal beds and in thick surface layers of organic materials in sub-Arctic tundra regions).

Currently, there is approximately 40% more carbon dioxide in the atmosphere than in the pre-industrial era. This additional carbon comes from layers that have been unreachable by living organisms for tens to hundreds of millions of years. The carbon-14 that existed in the organisms before they became compacted into fossil fuels has decayed to negligible levels over the millions of years during which the carbon has been hidden underground, out of reach of cosmic rays. The mixing of carbon from fossil fuels depleted of carbon-14 with atmospheric carbon in which carbon-14 continues to be created has caused a marked drop in the carbon-14 to carbon-12 ratio, which must be corrected for when using carbon dating over multiple millennia before the present.

A more recent complication in carbon-14 dating was caused by hundreds of nuclear explosions in the Earth's atmosphere and oceans conducted from 1945 through 1980. These explosions have added considerably to the carbon-14 content of the atmosphere: by 1964, the carbon-14 concentration in atmospheric carbon dioxide had nearly doubled because of the nuclear explosions conducted most frequently by the United States and the Soviet Union. This reservoir is gradually dissolving into the oceans and being taken up into the biosphere. By 2010, the residual effect of these tests amounted to an additional 10% carbon-14 in the atmosphere compared to pre-nuclear times. As every cell in a human body has, on average, a dozen carbon-14 atoms within it, a little more than one carbon-14 atom in each of our cells has not been created by cosmic rays smashing into the Earth's atmosphere, but instead by humans testing the devastating power of nuclear weapons. With some 4000 carbon-14 atoms radioactively decaying in our bodies every second, about 400 of those will represent the afterglow of the Cold

War on Earth, with the rest effectively being the afterglow of explosive stellar deaths far away in the Galaxy.

Key Points: Chapter 12

- During a total solar eclipse we see coronal rays projecting into inter-planetary space, which are shaped by the Sun's magnetic field and the solar wind.

- Electric currents heat the solar atmosphere to millions of degrees. This causes gas to evaporate off the Sun, forming the solar wind, which drags parts of the Sun's magnetic field with it.

- The Sun's magnetic field patterns reverse polarity on a quasi-regular 22-year cadence. The ever-changing solar dynamo powers explosions in the solar atmosphere and geomagnetic storms when their remnants flow past the Earth.

- When ejected solar magnetic field interacts with that of the Earth, atoms are accelerated that, on impact with the Earth's atmosphere, cause oxygen and nitrogen to glow in our skies, thus forming the auroras.

- The gusty, magnetized solar wind hampers the inflow of cosmic rays from the Galaxy. Variations in solar activity are consequently imprinted on the concentration of carbon-14 and other radioactive atoms that are created by the cosmic rays interacting with Earth's atmosphere, and that find their way into all life, including each human cell.

- **The solar dynamo triggers solar flares and coronal mass ejections, causes geomagnetic storms, and influences the Earth's magnetic field, leading to the northern and southern lights.**

But what exceeds all wonders, I have discovered four new planets [the moons of Jupiter] and observed their proper and particular motions, different among themselves and from the motions of all the other stars; and these new planets move about [Jupiter] like Venus and Mercury, and perchance the other known planets, move about the Sun. As soon as this tract, which I shall send to all the philosophers and mathematicians as an announcement, is finished, I shall send a copy to the Most Serene Grand Duke, together with an excellent spyglass, so that he can verify all these truths

GALILEO GALILEI (1564–1642),
in a letter to the Tuscan court from 1610,
quoted by Albert van Heiden,
Siderius Nuncius or The Sidereal Messenger (1989)

Scientists still do not appear to understand suficiently that all earth sciences must contribute evidence toward unveiling the state of our planet in earlier times, and that the truth of the matter can only be reached by combing all this evidence.

ALFRED LOTHAR WEGENER (1880–1930),
in *The Origins of Continents and Oceans* (1929)

13

Building a Home

Hydrogen dates back to the formation of the universe itself in the Big Bang. Nuclear fusion forged the other chemical elements out of which our planet and every living organism on it are made. This happened both deep within stars and during their explosive obliteration. In the final stages of a star's life, fused elements are ejected into the Galaxy, where they float around in interstellar space to mix with gases, some formed in the first minutes after the beginning of the universe and others ejected from other stars much more recently. Our planetary system was created out of such a mixture of gases that escaped from generations of stars that came before.

Until quite recently, we had no idea how many other planetary systems existed. In fact, we had no evidence of any other planets in the universe other than the Earth and its neighbors orbiting our Sun. It was not until the fall of 1995, in Florence, Italy, at an international meeting for astrophysicists working on solar and stellar magnetism, that the first detection of a planet orbiting a Sun-like star outside our own planetary system was reported. The news was announced in one of a whole series of presentations. Although there was appreciation for the work, not many among the attendees would have guessed how rapidly new planets would be discovered over the subsequent few decades, or would have anticipated just how many planets are by now estimated to exist in every galaxy in the universe.

The discovery of that first planet outside the solar system took resources then at the limit of the possible. How can something be detected that is at least dozens of times smaller in diameter than the star it orbits, and that reflects only a tiny fraction of the glow of that star? That first planet was detected indirectly, not by its own light, but by its slight pull on the massive star. When two or more objects orbit each other subject to their mutual gravitational attraction, they revolve about the common center of mass. For two objects of comparable mass, the center of mass lies halfway between them, and both objects orbit around that point. Suppose that we are looking at two stars that are orbiting each other, an arrangement that is, in fact, very common, because almost half of the bright points in the night sky visible to us are not single stars, but are actually binary systems (i.e. pairs of stars). For an observer at some distance in the plane of the orbits of such a binary system, the stars will move alternatingly toward and away from that observer.

In many ways, light behaves like waves, similar to waves of sound. That helps us determine whether the source of the light is moving toward or away from us. We know this works for sound waves. Sound from a car or from an airplane traveling toward us has a somewhat higher pitch than the sound of that same vehicle when it moves away from us. We refer to that change in pitch as the Doppler effect. Similarly, the light emitted from a star will appear to change in color, being a little bluer when the star approaches the observer and a little more red when the star moves away from the observer. If one star in a binary system is much brighter than the other, the light from the fainter one may not be discernible from the light of the brighter star. Yet, astronomers can detect the presence of the fainter component because the light from the brighter star drifts slightly from blue toward the red end of the spectrum, and then back toward blue as that star orbits its fainter companion. The less massive the companion, the closer the center of mass of the binary system will be to the heavier star. If the mass of the heavier star is very much larger, the center of mass of the pair may even lie within the heavier of the two. In that case, the velocity swing of the heavier component due to the pull from its much lighter counterpart is quite small as the star moves around the center of mass. Companions that weigh much less than the brighter, heavier star are therefore difficult to detect.

At that 1995 meeting in Florence, Michel Mayor presented the work that he and his colleague, Didier Queloz, both from Geneva, Switzerland, had been focusing on. They had patiently measured the light from 140 stars like the Sun over a period of 18 months. Their instrument unraveled the colors of the starlight to look for the Doppler effect, as that would be the telltale signature of the presence of faint, small companions to these stars. The very sensitive instrument they used enabled them to detect companions as light as one-thousandth the mass of the primary star. They found one of those very-low-mass companions to be so light that it could not be a star. It turned out that they had discovered the first planet outside the solar system. The planet weighed somewhere between half as much as our solar system's heavyweight planet,

Jupiter, to being comparable to Jupiter in mass. They could detect this planet not only because of their instrumentation and technique, but also because it orbited close to its star: this planet orbits its star in a mere 4.2 days, instead of 11.9 years as does our Sun's Jupiter, making the gravitational pull between the star and the new planet very much stronger. That discovery was greeted with quite some enthusiasm by their colleagues, but the impact it created in the world was remarkably muted. Demonstrating that the Earth, and indeed the entire solar system, is not a particularly special thing in the universe (apart from the fact that we live there) apparently did not shake a world in which aliens from all sorts of worlds outside of the solar system were already well embedded in the cultural psyche ever since the days of Jules Verne, television series such as *Star Trek*, and the myths about Area 51. Things were very different, however, way back when the Earth lost its place as the apparent center of the universe.

When, in the year of his death, in 1543, Nicolaus Copernicus published *De Revolutionibus Orbeum Coelestium* (*On the Revolutions of the Spheres of Heaven*) suggesting that not the Earth, but the Sun was at the center of the planetary system, this created some stir. Copernicus, anticipating trouble, had made it a highly technical book, in which he advocated his model as something of practical use but not necessarily as the physical truth. Consequently, the Church did not take official action against the book until 1616, when it was taken out of circulation pending modifications that would make it acceptable (after which the Church did not take it off the index of forbidden books in its original form until 1758). It was the publication of Galileo Galilei's *Siderius Nuncius* (*Starry Messenger*) in 1610 that caused the Church to react more aggressively to the new scientific discoveries. Galileo's observations with the newly invented telescope (to which he had made his own improvements) of moons orbiting Jupiter and of the phases of Venus suggested to him that the Sun was, indeed, the center of the planetary orbits. From then on he defended the position that the Sun was at the center of the planetary system, and not—as the Church held to be true—the Earth. That led the Church to take action against him. Not long after that, Isaac Newton devised the mathematical formulation for gravity and resulting orbits. He realized that not even the Sun was at the center of the planetary system, writing in his *Philosophiae Naturalis Principia Mathematica* in 1687 that (translated from the original Latin by Benjamin Motte in 1729) "the common centre of gravity of the Earth, the Sun and all the Planets is to be esteem'd the Centre of the World" (where the word "World" is used to describe the solar system).

It was the effect of planets tugging stars around precisely such a "common centre of gravity" that allowed the discovery of the first planets outside the solar system, known as exoplanets. After the initial discovery in 1995, specialized instruments placed in space proved to be so sensitive that they did not need to look for the effects of that gravitational tugging, but could even measure the slight decrease in a star's brightness when an orbiting planet crossed in front of it. Measuring how fast a star's light decreases when a planet moves in front of it and increases again when it moves on enables researchers to measure the diameter of the planet, by combining the law of gravity with knowledge of the star's dimensions. By combining both methods, the Doppler measurements of the gravitational pull and the measurements of the blocked starlight on planetary transits, even the mass of exoplanets can be estimated.

The first exoplanet to be discovered orbits 51 Pegasi, which is a star quite similar to the Sun. The planet, referred to as 51 Pegasi b, is not at all friendly to life, however. Its proximity to its sun causes its average surface temperature to be some 1200 degrees centigrade.

Following the first discovery, the rapid advance in technologies has led to the confirmed identification of close to 1000 planets, with many thousands of other measurements producing good candidates for planet confirmation. Even the nearest Sun-like star, Alpha Centauri B, has been shown to very likely have at least one very hot planet.

Finding small planets like the Earth at a sufficiently large distance from a star to make its surface temperature comparable to that of the Earth is very, very challenging. The gravitational pull of such a small, distant planet on the central star is minute, and the size of the planet is tiny compared to that of the star. Our Earth, for example, would fit inside the Sun 1 million times. Because of this contrast in sizes and masses between star and planet, both the Doppler signal and the transit darkening are very difficult to measure. So, we do not yet have a good handle on how many planets there really are out there in our Galaxy. Estimates of the average number of planets per star differ somewhat, but they suggest that at least one in five Sun-like stars has a Jupiter-like large planet and at least two in five have smaller planets. One study concluded that stars are orbited by one or more planets as a rule. The consequences of that are simply mindboggling: with several hundred billion stars estimated to exist within our own Milky Way alone, it means that there is a similar number of planets out there to be discovered, studied, and—at least for the more favorably located ones—considered as potentially supporting life.

As yet, no life has been discovered on planets other than our Earth. We have much to learn about what makes planets at least in principle habitable, but one criterion is commonly used to select those planets: the planet should be able to have liquid water on its surface. With this criterion, perhaps 1 in 100 planets could, in principle, be habitable. Given the number of planets that we now think exist in the Galaxy, this leaves us with an expected 1 billion planets or so that we should consider as potentially habitable.

Looking at other stars and their planets can help us to understand how our own solar system formed. After all, stars can be seen in all stages of their development somewhere in the Galaxy, from even before their formation (known as stellar birth) to after their ultimate explosion (or death) or—if their mass is too low to experience an explosive end—in their late phases as compact post-stellar objects merely glowing ever more faintly, possibly until the end of time. This method is in some sense not dissimilar to learning about human aging, for example. By looking at a reasonably large sample of people on any one given day we can learn a lot about aging. Although we do not see any individual actually grow appreciably older within such a short period, we would see the full spectrum of ages and could piece together a fairly complete picture of the human life span, just from this single day of observing. With stars, it is even easier: the laws of physics map out for us how to connect one stage to the next in remarkable detail when they are combined with such observations.

How do these planetary systems form? Present-day computers are challenged to their limits to find out how the central stars and their orbiting planets form and subsequently evolve. What they have produced, though, is at least a rough storyboard for all of the atoms that make up our world, as well as our bodies. At the very beginning of a planetary system there is a vast cloud of cold gas, in which ancient hydrogen and helium are mixed with heavier elements that were thrown off earlier generations of stars during the pulsations or explosions they undergo when they run out of fuel for their nuclear furnaces. Gravitational ripples from the surrounding galaxy, or reverberating waves from distant stellar explosions, may make some parts of these vast clouds somewhat denser than others. If there is enough mass in a given cloud core to have its own gravitational pull win over the outward pressure of the gas, it can begin a long process of compression toward what will eventually be the star at its center. Astrophysical laws reveal that for a cloud to condense into a star, its size has to exceed at least some 20,000 times the Sun–Earth distance. The motions within such vast clouds, slow whirls and eddies subject to the pull of multitudes of distant stars, invariably have some net angular momentum. In the oft-used analogy of the ice dancer who spins up when she pulls in her arms, the contracting cloud will spin ever faster as it becomes more compact under the influence of its own gravity. As stars are at least a million times smaller than the original cloud out of which they form, this presents a formidable problem to the formation of a star.

While contracting, the cloud would spin faster and faster, until at some point the outward centrifugal forces would be comparable to the inward pull of gravity and the cloud would simply no longer be able to contract. This is true in the direction of the spinning, but not in the direction perpendicular to it. The result is that the cloud can, at first, collapse into a revolving gaseous pancake, that is, a disk. Large telescopes sensitive to the faint red glow of these collapsing clouds have confirmed the existence of clouds of the required size prior to collapse, and also reveal the existence of remnants of disks around the youngest stars. The problem of how to move the rotational energy out of a revolving disk to ultimately enable the formation of a star continues to be studied. Magnetic fields, stellar winds, and gravitational pull all appear to play important roles in some as yet unexplained concerted action. With the multitude of planets now discovered, it is clear that some fraction of the rotational energy of the cloud is still within the planetary systems, contained in the planets' orbital motions about the central stars. This scenario is now not merely a hypothetical way to let a star be formed, but appears a ubiquitous, real process leading to vast numbers of planetary systems.

The initial collapse phase for the cloud out of which our solar system formed would have lasted on the order of 100,000 years. Ultimately, probably growing in spurts punctuated by periods of slow evolution, the young Sun would have gathered its mass. Somehow it did not split into a pair of stars, although that is what happens in roughly half of all the forming stellar systems, causing them to end up as binary stars. Our Sun remained intact and therefore single for all time. At some point, about a million years into its formation process, the young Sun transitioned from a glowing ball of warm gas falling in on itself into a star proper: a sphere of predominantly hydrogen and helium in which gas pressure just balances gravity, and in whose core nuclear reactions supply the energy to keep the gas hot and its pressure high even as energy leaks away from the surface in the form of starlight.

Stars form with vastly different masses, from somewhat more than 100 times the mass of the Sun down to just under one-tenth, in other words over a full range of more than a factor of 1000. This is not to say that even lighter objects do not form but these are not, technically, stars. For objects with too little mass, the internal temperature and pressure are insufficient to initiate nuclear fusion of hydrogen. In that case, the entity is known as a "brown dwarf" rather than as a true star. These brown dwarfs were unambiguously confirmed to exist only in the mid-1990s, after the first somewhat uncertain detection in 1988.

Brown dwarfs do have some limited nuclear fusion going in their interior, but that involves deuterium (the "heavy hydrogen", with an extra neutron in the nucleus) rather than pure hydrogen. There is but little deuterium in stars, however, so the energy that is released by its fusion is limited. If no nuclear fusion occurs at all, which is the case when the object has a mass of less than approximately one-eightieth of that of the Sun, it is formally designated a planet. How many brown dwarfs or free-floating planets form is presently unclear, because they are very faint objects glowing mostly deep red, rather than as bright white or yellow stars. As a consequence, they are very difficult to find even with large telescopes. These relatively small objects, too small to be true stars, may themselves have miniature planetary systems, just as the large planets Jupiter and Saturn have their moons within our own solar system, but we have yet to discover whether such constellations do indeed form.

Very young stars and their nascent planetary systems are seen in many directions in the sky and their observations help us assemble the history of the young Sun with a fair degree of certainty. Two properties make very young stars fundamentally different from the current Sun. The first is that these young stars are covered by starspots, the stellar analogs of the magnetic sunspots, which are very much larger and more numerous than sunspots are on the present-day Sun. The record holders have spots that cover half their surface. The primary reason for this large starspot coverage is that a young star spins very quickly, typically with a rotation period of less than a day. In contrast, the present-day

Sun has an average rotation period of 25 days. Over time, the spin rate slows down because the strong magnetism drives a forceful stellar wind that carries rotational energy away from the star. This causes an associated reduction in starspot coverage over the billions of years of maturity for a Sun-like star.

The second distinction between young and mature stars is that young stars still have the remnants of a disk of gas and dust around them. At first, stars accumulate their matter from the disk as they are growing. That process, called accretion, gradually tapers off over a period of just a few million years. At the age of a million years, almost all forming stars have disks, whereas at five million years only 1 in 10 has a measurable disk of dust and gas left. Far from a star, the forces of the orbiting gas exceed those of the star's magnetic field, while close to the star the stellar magnetic field wins and the gas in the disk cannot move freely in its normal gravitational orbit. Instead, the magnetic field takes the gas from the innermost edge of the disk, where the stellar field begins to dominate, and guides it along that field to accrete wherever that field connects to the stellar surface. The impacting material becomes very hot, and its glow can then be observed, revealing strong fluctuations that suggest that sometimes as much as 30 Earth masses per year can fall onto a young star for periods of at least a century, with much less accumulating at other times.

In the meantime, the micron-sized dust particles in the vast disk that slowly whirls around the forming star—as it would have in the case of the solar system—grow bigger as particles collide and stick to each other. Within some millions to a few tens of millions of years, these particles become large enough for the disk to be no longer so generally filled with gas. Thus, instead of being opaque, the disk becomes transparent. This transition happens first closer to the star, and progressively later further away from the star. As we now know from the abundance of stars with planetary systems, this is a common process throughout the Galaxy, and therefore likely to occur in any of the billions of other galaxies that we can see around our own.

Even as the dust grows into larger chunks through the accretion process, the lightest residual gases, hydrogen and helium, are mostly blown out of the disk by a combination of radiation pressure and stellar wind. This results in a pre-planetary disk composed of most of the chemicals in the interstellar medium and the star that formed out of it, except for the amounts of those two most abundant and most volatile of gases that were blown out of the system during its formation. Exactly when and where a planet forms relative to the time when hydrogen and helium are blown out of the area of the disk likely determines whether the planet becomes a gas giant (such as Jupiter, Saturn, Uranus, or Neptune) or a rocky "terrestrial" planet (like Mercury, Venus, the Earth, or Mars).

One scenario for the formation of the Sun's planetary system has micron-sized dust form 10,000 years after the initial cloud began its contraction, reaching kilometer-sized chunks by some 30,000 years, which then eventually would grow to full-sized planets after tens of millions of years. After that initial formation, major collisions between pre-planetary bodies of various sizes will continue for many tens of millions of years more. Each of such powerful collisions might be energetic enough to melt much of the growing planet. One was at least strong enough to make the proto-Earth shatter into pieces that would later reconfigure into two big bodies: the current Earth and its Moon.

With things settling down in the early solar system after some 600 million years, the Earth's history shifted from its basic formation to terraforming, as its ever-evolving life forms came into existence. Some residuals of the basic formation process are still around. Far away, past the outer planet, Neptune, and more than 10,000 times the Sun–Earth distance, is the Oort cloud, comprising many billions of sizable chunks of the most ancient materials in the solar system. Somewhat closer is a flatter cloud, called the Kuiper belt. A remnant of the formation of the solar system that is much closer to the Earth than the Oort cloud or the Kuiper belt is the ensemble of minor planets. These, better known as asteroids (from the Latin for "star-like"), are rocky bodies up to about 1000 kilometers or 600 miles across. The vast majority of the asteroids orbit the Sun in a belt between the orbits of Mars and Jupiter. There may be as many as 1–2 million that are larger than a kilometer in cross-section, and many, many more smaller ones.

In the latter half of the eighteenth century, Johann Daniel Titius and Johann Elert Bode, both German astronomers, commented on an algorithmic regularity of the distances from the Sun of the six planets then known. By 1783 a seventh planet was discovered. Bode suggested the name Uranus, and noticed that it, too, adhered to the Titius–Bode law of planetary distance. There was, however, a curious gap between the fourth planet, Mars, and the fifth planet, Jupiter, where the Titius–Bode law would have required another planet to exist. This "law" was a mere observed regularity in a pattern, rather than a tested law of physics. Yet, it had created enough interest for astronomers to hunt for what appeared to be the missing planet. By 1801 it looked like this missing planet had been found: Guiseppe Piazzi, the director of the observatory in Palermo, Italy, cautiously reported on the discovery of what he, at first, named a potential comet. Establishing something as an orbiting planet required more observations than Piazzi could make in 1801.

With help from Carl Friedrich Gauss (of the Gauss curve, among other things), who computed the orbit based on Piazzi's initial measurements, the object was found again at the end of the year. It was established to be pretty much right where the Titius–Bode law predicted a planet to be. Piazzi's initially proposed double name was later shortened to the first part: Ceres. It proved not to be a full-sized planet, however: its diameter of 950 kilometers (not even a third of the diameter of our Moon) is large enough for its gravity to have made it into a sphere, like the other planets, but it is officially designated a dwarf planet by the International Astronomical Union. The reason for that designation is that Ceres is not heavy enough to have captured the multitude of other objects close to its orbit. This was realized within six years of Ceres' discovery, when other rocky objects (Pallas, Juno, and Vesta) were identified not too far from Ceres' orbit. As telescopes grew more powerful and as computational methods increased in accuracy, more and more asteroids were discovered. Most asteroids exist in the region between Mars and Jupiter, where the Titius–Bode law suggested another planet should exist. When Neptune was discovered in 1846, however, it did not adhere to the Titius–Bode law. We now realize that this "law" that triggered the hunt that led to the discovery of Ceres was not a physically significant rule after all, but a chance pattern.

A few million asteroids exist within the solar system out to the orbit of Jupiter alone. Most occur in the asteroid belt between Mars and Jupiter. They range in size from the nearly 1000 kilometers of Ceres, to rocks too small to detect from the Earth. Perhaps 10,000 of them are larger than about 10 kilometers (6 miles) across, a million or so larger than 1 kilometer, some 25 million larger than roughly 100 meters (yards), and beyond that we simply do not know. There are many, but they do not contain enough mass to make a proper planet, not even all of them combined. An estimate of the total mass of all the asteroids of our solar system is approximately 0.05% of that of the Earth, and a third of that resides in the largest one, Ceres.

Not all asteroids, however, move around the Sun beyond Mars, and at a safe distance from our planet: there are close to about 1000 asteroids larger than 1 kilometer in diameter that orbit the Sun at a distance comparable to that of the Earth, and many thousands more that are smaller. Asteroids in orbits like these sometimes hit the Earth. Over time, these collisions clear out the near-Earth objects, but the gravitational pull of the major planets perturbs the orbits of the asteroids between Mars and Jupiter, replenishing those lost in collisions. On average, one asteroid with a diameter of up to 10 meters (yards) hits our atmosphere each year; these typically explode high in the stratosphere, with little effect at the Earth's surface. Once every few millennia, an object with a diameter of some 50 meters collides with the Earth. One such collision happened in 1908, near the Podkamennaya Tunguska River in Russia. That "Tunguska event" had an estimated strength of several tens of millions of tons of TNT. The explosion is estimated to have knocked over approximately 80 million trees in the central Siberian region and would have killed all life in its path. Catastrophic collisions with kilometer-sized objects or larger happen much less frequently, perhaps once every million years. There are now ongoing astronomical projects to identify all near-Earth objects, so that we shall at least know that a big one is coming well ahead of time, hopefully well enough ahead of a potential collision to develop the means to prevent it.

Asteroid and comet impacts happen on any body within the solar system, even to asteroids and comets themselves, leaving behind craters on their surfaces if they are not shattered in the process. Impacts that are sufficiently large will cause material from the impacted body to be ejected into space, from where it may travel to another body. In 1982, the mineral analysis of an unusual-looking meteorite found in Antarctica demonstrated that it was very similar to that of rocks that had been brought back from the Moon by the Apollo astronauts. In the years since then, dozens of meteorites have been identified that originated from the Moon. Rock fragments that left the Moon after a meteorite impact with a velocity slightly larger than the Lunar escape velocity can find their way to the Earth and may impact within years, but others may take thousands of years.

Determining the times it took for the Lunar rocks to reach the Earth, as well as how long they have been on the Earth, relies on the analysis of radioactive atoms within these rocks that are created when cosmic rays impact them. Buried rock on the Moon is not exposed to cosmic rays.

When traveling through space after being thrown off the Moon by a meteorite impact, however, all sides of the rock are exposed. Once such a rock falls onto the Earth, thereafter lying still, the upper and lower sides are exposed differently, so that the analysis of distinct isotopes with different decay times can reveal these travel and residence times.

Lunar meteorites make up approximately 1 in 300 of all found meteorites. Some meteorites have been identified as having originated from Mars. In the early 1980s, several such identifications were made that convinced the scientific community: the chemical analysis of small pockets of gas trapped within the meteorites showed the gas to have the same chemical composition as the atmosphere of Mars. The known Martian meteorites were ejected between 1 and 20 million years ago. Only a few dozen meteorite fragments are known to have originated from Mars. The total mass of identified Lunar and Martian meteorites is minute compared with the mass of the Earth. In earlier ages, when the frequency of meteorite impacts was much larger, and over the billions of years in between, much more material will have been exchanged between the Moon, Mars, and the Earth than now identified. It is therefore likely that somewhere in our bodies traces of that exchanged material exist. The most likely candidates for this origin are oxygen and iron, because these are the most abundant elements in such meteorites, and are also among the most plentiful within the human body.

Key Points: Chapter 13

- There are close to 1000 confirmed planets outside our solar system, but there may be several hundred billion within our Galaxy, about the same number as the number of stars. In our Galaxy alone, about 1 billion planets may have surface water and be habitable.

- Planetary systems originate from ancient clouds of interstellar gas. If there is enough mass in a cloud for its gravitational pull to overcome the outward gas pressure, it begins to collapse toward what becomes its central star.

- Stars have vastly different masses. Objects too light for hydrogen fusion are not stars but brown dwarfs. Planets are objects with no nuclear fusion at all.

- A large number of small bodies, called asteroids, orbit the Sun. The vast majority of them exist in a belt between the orbits of Mars and

Jupiter. Those pulled from their orbits by the larger planets occasionally hit the Earth.

- Early in the history of the solar system, pre-planetary bodies collided. One such collision broke the proto-Earth into two pieces: the Earth and its Moon. Fragments of such collisions have landed on the Earth, so that our bodies contain traces of other planets, asteroids, and comets.

- **Many stars have a planetary system similar to the solar system and their process of origination and development is similar to ours.**

All birds which fly have round their leg the thread of the infinite. Germination is complicated with the bursting forth of a meteor and with the peck of a swallow cracking its egg, and it places on one level the birth of an earth-worm and the advent of Socrates. Where the telescope ends, the microscope begins. Which of the two possesses the larger field of vision?

Victor Hugo (1802–85),
in Les Misérables (1862)

14

Stardust in Flux

Insight into what our bodies are, and what they are not, jolts our view of ourselves into a new perspective. We all acknowledge that our bodies change as we transition from birth to adulthood. After that, we tend to think of ourselves as lasting organisms that take in the chemical nutrition of solids and fluids, and inhale air that powers the body through chemical reactions. Most of that concept is true enough, but we do not last very long without fundamentally exchanging with the world around us the materials of which we are composed.

We exchange most of what makes us—water—in a matter of at most weeks. The structures of our cells, and even the cells themselves, decay to be replaced on time scales from weeks to a few years, or at most a few decades for some organs. Even our bones are subject to replacement as the calcium and other atoms are replaced, and we leave our DNA on almost everything we touch. The only lasting parts of us appear to be our teeth, which are, ironically, not really alive at all: they are deposits of lifeless material that serve us well, sometimes throughout our lives, even as the rest of our bodies is replaced over time. Consequently, the saying that we are not the same who we were before is far more apt than we generally realize. Science has yet to uncover whether the memories stored in our bodies are subject to the same replacement. Some of our brain cells do survive for long times, even if the materials out of which they are made are replaced, but it appears that we have to acknowledge that it is more likely that even memories are patterns that are exchanged rather than particular fixed atomic clusters that remain with us from the moment they are formed to the moment we forget, or die. All of this happens throughout our lifetime, but usually we are oblivious to all this action that takes place in the background of our daily lives. We often become acutely aware, however, when our body "malfunctions", during periods of illness. Pathologists, who are the medical specialists who diagnose diseases and their causes, also study the responses of the body to such disease processes and to the treatment thereof. The insight of the profoundly transient nature of our bodies is quite a concept to take in.

When we trace the pathways of our impermanence, we realize that the components of our bodies connect us to the plants and animals around us, bacteria within us, to volcanism, comets, cosmic rays, and to the Sun's light, all the way to the birth and death throes of stars throughout the Galaxy and to the beginning of the universe itself. The most common element in our bodies, hydrogen, is as old as the universe. Other elements were formed generally over 5 billion years ago, in the interior of stars. Some atoms were formed later, as radioactive stardust within the Earth, which decayed to create, for example, some of the calcium in our bones. A little of the carbon in everything we eat was formed within only thousands of years, created somewhere in the Earth's stratosphere when shards of atoms from far away, having bounced around the Galaxy for millions of years following an initial kick from a stellar explosion, finally ran into our planet. A smaller fraction of all the carbon around and inside us was manmade, resulting from atmospheric nuclear tests just a few decades ago. Other atoms started out on the Moon, on Mars, in asteroids and comets, or elsewhere in the solar system, and fell to the Earth sometime between the birth of the planet and days ago. All of that material cycles through bacteria, plants, insects, reptiles, birds and mammals, capturing sunlight that was created many thousands of years ago in the nuclear furnace in the core of the Sun. Everything we are made of also moves through the depths of the Earth, in the geological cycles associated with continental drift. That deep-Earth cycle, as well as the use of fossil fuels, connects the materials that build us to extinct mammoths and dinosaurs, and even to the first life on Earth.

Understanding what constitutes our bodies weaves a story through the complexity of organic chemistry and the identification of fossils, through developing nuclear physics and the discovery of DNA, via the origins of the elements, and cell growth and division. The knowledge needed to understand any particular problem that scientists run into often comes from apparently disparate disciplines. Nature does not respect any of the human attempts to cleanly separate the natural sciences into independent research fields, so that insights often develop across the interfaces between two or more disciplines, and sometimes entirely new fields of science have to be created. Moreover, scientists delve ever deeper, creating specialties within sub-disciplines: each of the chapters in this book would be readily expanded into multiple volumes, with each level revealing more detail and shifting the frontline of our questions beyond the current battles in the minds of those engaged in furthering humanity's knowledge.

With the rapid increase in our insights into nature's workings, and with the associated increase in specialization and the fragmentation of scientific disciplines, it has become impossible for anyone to maintain an up-to-date knowledge of it all. Up to some three centuries ago it was possible to be very knowledgeable about much of what was then known about science. Such a learned person was known as an *uomo universale*, a polymath, a universal man (and, indeed, at the time most were men). The most accomplished of these thinkers were multifaceted in their approach and explored as much as they could in both the arts and the sciences. In the nineteenth century, science expanded considerably, yet some bright minds maintained a rather broad view and certainly a wide-casting curiosity. These days, some such multidimensional people still exist, but the vastness of our knowledge makes it a fantastic challenge to successfully work across what are by now highly differentiated sciences. At the end of the sixteenth century, we can see Galileo Galilei working on military instruments to improve the use of guns, strengthen the magnification of the spyglass so he could discover moons of Jupiter and study spots on the Sun, and study gravity and tides on the side. In the seventeenth century, we could still encounter a Robert Hooke, who worked as a London city surveyor while designing telescopes, discovering cells in biological tissue, and pondering paleontology. Even in the nineteenth century, we can think of Rudolph Virchow, who argued that the cell is the basic unit of health and disease, and who is now known as the father of cellular pathology. Apart from his medical endeavors, he studied languages, archeology, and anthropology, was a terrific teacher, and led a successful personal life.

Quickly, though, science grew too much for it all to be kept in view. Consequently, scientific advances nowadays are more often the result of community efforts, in which individual contributions are typically hard to single out (consequently, we did not attempt to do so in this book, with few exceptions, for those discoveries made in the past century). Yet, there are many examples of leaps in understanding that stand out because of scientists' awareness of developments beyond the bounds of their own discipline. Science often delivers its largest advances when two or more of its disciplines are forced together to solve a problem.

In any given field, however, such leaps may happen only sporadically: some of the issues discussed in this book saw decades or centuries pass before perspectives and insights shifted dramatically. Interestingly, each leap forward creates new questions and only increases our sense of wonder, scientific curiosity, and fascination with "our world".

The links discussed in the story of our bodies are not only philosophical connections between the sciences, but involve physical connections to what we consider to be nature outside us. For one thing, the cells in our bodies are outnumbered by approximately 10 to 1 by the microbiome in our intestines and on our skin. In the microbiome that we host, several hundred species of bacteria mixed with other micro-organisms are in a reciprocal relationship, where we benefit from the activities of the bacteria, for example in the overall digestive chemical cycles, while they benefit from being in a relatively benign, stable environment, where nutrients and water are supplied by the actions of their hosts. The interdependence of life forms goes well beyond the body, of course. We, as animals, depend on the ability of plants to capture sunlight in order to produce a wide range of chemical compounds, while in the process liberating oxygen. We need both of those categories of chemical products to be able to live. Moreover, plant growth is an important mechanism by which carbon dioxide in the atmosphere is regulated: carbon dioxide regulation by plants is an essential link in the chain of processes that determines the diversity of climates that provide suitable living conditions for us and for the plants that we consume.

The story of our bodies is one of impermanence as much as one of connections. All of us are universes unto ourselves, and yet, at the same time, frail organisms that survive only because we are cradled by a network of other life forms. In our story, we focused mostly on the structure of our bodies and on the physical processes involved in how materials that contribute to it cycle through the world around us. That story is woven around the workings of the ecosystem in which we live, which has its own myriad connections. We did not go into many of those, except for the few links needed to trace some chemicals, in particular where these concerned photosynthesis and the nitrogen cycle. The mesh of life forms involved in enabling all of life to survive, as studied by ecologists, is but one view within the perspective of the story we told here. Yet, that zoomed-in view, too, presents an immense richness, filled with surprises and many unsolved mysteries. We chose to focus on the flow of elements and simple compounds, hardly touching on how those flows themselves were sustained: the lives of animals, plants, fungi, bacteria, and archaea are intertwined and interdependent in so many ways that we are continually surprised to see the consequences of our actions ripple through the world around us. We would do well to treasure the labyrinth of life forms that supports us, to ensure survival for ourselves and for our beautiful blue planet. The planet-wide ecology of the human species contains a treasure trove of surprising links beyond the few that we could describe here, but one thing should be clear: nature is not just out there to be looked at or exploited by us as unattached bystanders, but is inextricably linked to our very well-being and survival.

Index